Systematics and the Exploration of Life

Series Editor
Marie-Christine Maurel

Systematics and the Exploration of Life

Edited by

Philippe Grandcolas
Marie-Christine Maurel

iSTE

WILEY

First published 2021 in Great Britain and the United States by ISTE Ltd and John Wiley & Sons, Inc.

ISTE Ltd
27-37 St George's Road
London SW19 4EU
UK

www.iste.co.uk

John Wiley & Sons, Inc.
111 River Street
Hoboken, NJ 07030
USA

www.wiley.com

Library of Congress Control Number: 2020949898

British Library Cataloguing-in-Publication Data
A CIP record for this book is available from the British Library
ISBN 978-1-78630-265-6

MIX
Paper from responsible sources
FSC® C013604
www.fsc.org

Contents

Introduction

Exploring Biodiversity: Science Must Seize the Unknown 80%

We actually know far too little about biodiversity! We are idly living on the improved achievements of a period of intense exploration, which lasted from the 18th century through to the beginning of the 20th century, with the beginnings of "systematics".

The modern formalized description of the diversity of life was born at the beginning of this period, namely the famous *Systema Naturae* by Carl Linnaeus (1758). These first classifications were constructed on the basis of an implicit order in life, as perceived by precursor authors. This comparative perception and the linking of structures between different organisms are, indeed, at the heart of the origin of the theory of evolution (Le Guyader 2018; Montévil 2019).

While the progress of systematics waned at the beginning of the 20th century, general biology developed extraordinarily. It focused on the study of the laws of life through the study of a few organisms that imposed themselves as "models", from the vinegar fly to the white rat. Immense discoveries were made about heredity, the functioning of organisms and living cells, which today form the basis of our general knowledge (Mayr 1982). In comparison, the exploratory and still descriptive approach to the diversity of living organisms was gradually becoming obsolete; it suffered from enunciating particulars rather than the universals of general biology (Mahner and Bunge 1997; Grandcolas 2017).

Introduction written by Philippe GRANDCOLAS and Marie-Christine MAUREL.

Fortunately, the subsequent development of a comparative methodology and phylogenetic analysis revived this field and enabled it to make a strong contribution to modern evolutionary biology (Nelson 1970). Biology then rediscovered the diversity of organisms (Wilson 1988), making a new synthesis by considering the general laws of life *and* the diversity of their expression in living organisms (Grandcolas 2018).

The balance sheet of these past decades of exploration is both extraordinary, commensurate with biological diversity and these glorious periods of discovery *ab nihilo*, and disappointing, as we too often capitalize on a false feeling of déjà vu (Grandcolas 2017).

And yet, to give just one figure, we currently only know about two million living species, in other words, less than 20% of the 10 million species whose existence has been statistically inferred on numerous occasions (May 1988). Study after study on the many groups of organisms shows how much remains to be discovered, whether small or large, or near or far from us (e.g. Bouchet 2006; Vieites *et al.* 2009; De Vargas *et al.* 2015; Hawksworth and Lücking 2017; Nicolas *et al.* 2017). We still know very little about most of the so-called "species known to science". We only have a few lines in old publications which describe more than half of the macro-organisms (Troudet *et al.* 2017) and a few molecules, instead of whole phenotypes on uncultured microbes (Konstantinidis *et al.* 2017).

The issue now is to understand that it is essential to discover the unknown 80% of biodiversity for several well-defined scientific reasons, more than for the thirst for new knowledge or for a compulsive collection of new species.

First of all, the laws of life have rather varied degrees of generality; from heredity to the functioning of ecosystems, for example, there are several orders of magnitude of difference in this respect! Many laws or principles require the study of more particulars in order to reach generality, given the variation that is the intrinsic property of living things (Montévil *et al.* 2016). Clearly, we need to know about more organisms and the particular cases of their biology in order to be able to claim to generalize. The rules of representativeness, dominance or abundance, stated as truisms, are often misleading in living organisms. For example, it has recently been documented that rare and scarce species often fulfill disproportionately important functional roles within ecosystems (Mouillot *et al.* 2013).

In contrast to universals, particulars very frequently remind us of how many pathogenic, invasive species are discovered in this way, having already crossed half the planet, causing us great concern. This is as much the case for HIV (Barré-Sinoussi *et al.* 1983), an obscure retrovirus from an African primate, as for an invasive and unknown flatworm imported from South America, threatening the fauna of our soils and their faunal balance (Justine *et al.* 2020). Not a week goes by without a species new to science presenting a question to our societies. The COVID-19 pandemic is a dramatic demonstration of this: here again, a few poorly known bats and pangolins harbor unknown (and described for the occasion) coronaviruses whose genetic recombination is putting the human world at a standstill (Hassanin *et al.* 2020). This is also the case for viruses and viroids of plants that are still largely unknown, and vectors of devastation in some plantations (Maurel 2018).

The issue of bio-inspiration (Benuys 1997) is another opportunity to understand how much the diversity of living things contains wonders from which we can draw inspiration for more sustainable societies; so many particulars (structures, functions, etc.) in different species whose natural function can be transposed to functions of human interest. Practicing bio-inspiration beyond random discoveries of opportunity requires a broad and reasoned exploration of living things and the relationships between their structures and functions.

Particulars are also often geographical rather than purely taxonomic. Each state, government or municipality needs to be aware of local biodiversity in order to develop a reserve or environmental, agricultural or health policies (Pellens and Grandcolas 2016). These are all reasons to be aware of local fauna and flora with their innumerable numbers of endemic species (Caesar *et al.* 2017). It is worth remembering the order of magnitude of these numbers and that there are, for example, 40,000 species of insects in metropolitan France alone (Gargominy *et al.* 2014).

Even if we focus on a few species for reasons of immediate interest, it is essential to know their close relatives. Knowing the meaning, adaptive character and selection regime of the traits of organisms, whether they are genotypic or phenotypic models, requires an understanding of their history (Jenner 2006). Is it necessary, once again, to quote Dobzhansky (1973) – "nothing in biology makes sense except in the light of evolution" – to be convinced of this. Reconstructing the origin and evolution of the traits of an organism of interest requires knowing not only its close relatives, but also a very large part of the living world. How many fundamental traits has the human species inherited, the understanding of which is based on their structure and function at the Metazoan scale (more than a million species!)? This presupposes an adequate taxonomic sampling of life, which is not necessarily limited to known species, but which must be searched for out in the field in order to find unknown species whose lifestyles have sometimes been long surmised.

The entirety of this book is therefore dedicated to these approaches to exploring the diversity of life, each of them showing the crucial need we have for exploratory approaches. "Exploratory", which is easy to understand when reading this volume, does not refer to a kilometric description of specific characteristics, but to an organization of knowledge and hypothesis tests, based on a large sampling of living species – a large part of which is, strangely, still unknown to us, even though we come into contact with it every day. Without further delay, we must not suffer from or destroy biodiversity, but study it in order to integrate it sustainably into our societies.

References

Barré-Sinoussi, F., Chermann, J.C., Rey, F., Nugeyre, M.T., Chamaret, S., Gruest, J., Dauguet, C., Axler-Blin, C., Vezinet-Brun, F., Rouzioux, C., Rozenbaum, W., and Montagnier, L. (1983). Isolation of a T-lymphotropic retrovirus from a patient atrisk for acquired immune deficiency syndrome (AIDS). *Science*, 220, 868–871.

Benyus, J.M. (1997). *Biomimicry: Innovation Inspired by Nature*. Harper Perennial, New York.

Bouchet, P. (2006). The magnitude of marine biodiversity. In *The Exploration of Marine Biodiversity: Scientific and Technological Challenges*, Duarte, C.M. (ed.). Fundacion BBVA, Bilbao, Spain, 31–62.

Caesar, M., Grandcolas, P., and Pellens, R. (2017). Outstanding micro-endemism in New Caledonia: More than one out of ten animal species have a very restricted distribution range. *PLOS One*, 12(7), e0181437.

Dobzhansky, T. (1973). Nothing in biology makes sense except in the light of evolution. *The American Biology Teacher*, 35, 125–129.

Gargominy, O., Tercerie, S., Régnier, C., Ramage, T., Schoelinck, C., Dupont, P., and Poncet, L. (2014). TAXREF v8.0, référentiel taxonomique pour la France : méthodologie, mise en œuvre et diffusion. SPN report, 42.

Grandcolas, P. (2017). Loosing the connection between the observation and the specimen: A by-product of the digital era or a trend inherited from general biology? *Bionomina*, 12, 57–62.

Grandcolas, P. (2018). The view of systematics on biodiversity. In *Biodiversity and Evolution*, Grandcolas, P. and Maurel, M.-C. (eds). ISTE Ltd, London and John Wiley & Sons, New York, 29–38.

Hassanin A., Grandcolas P. & Veron G. (2020). Covid-19: Natural or anthropic origin? *Mammalia*, 000010151520200044.

Hawksworth D.L. and Lücking R. (2017). Fungal Diversity Revisited: 2.2 to 3.8 Million Species. In *The Fungal Kingdom*, Heitman J, Howlett BJ, Crous PW, Stukenbrock EH, James TY and Gow NAR (eds). American Society for Microbiology, Washington, DC, 79–95.

Jenner, R.A. (2006). Unburdening evo-devo: Ancestral attractions, model organisms, and basal baloney. *Development Genes and Evolution*, 216, 385–394.

Justine, J.L., Winsor, L., Gey, D., Gros, P., and Thévenot, J. (2020). *Obama chez moi*! The invasion of metropolitan France by the land planarian *Obama nungara* (Platyhelminthes, Geoplanidae). *PeerJ*, 8, e8385.

Konstantinidis, K.T., Rosselló-Móra, R., and Amann, R. (2017). Uncultivated microbes in need of their own taxonomy. *The ISME Journal*, 11, 2399–2406.

Le Guyader, H. (2018). *Classification et évolution*. Le Pommier, Paris.

Linnaeus, C. (1758). *Systema Naturae per regna tria naturae, secundum classes, ordines, genera, species, cum characteribus, differentiis, synonymis, locis*, 10th edition, volume 1. Laurentii Salvii, Holmiae, Stockholm.

Mahner, M. and Bunge, M. (1997). *Foundations of Biophilosophy*. Springer, Berlin.

Maurel, M.-C. (2018). À la frontière du vivant: les viroïdes [Online]. *The Conversation*. Available: https://theconversation.com/a-la-frontiere-du-vivant-les-viro-des-90500.

May, R.M. (1988). How many species are there on earth? *Science*, 241(4872), 1441–1449.

Mayr, E. (1982). *The Growth of Biological Thought: Diversity, Evolution, and Inheritance*. Harvard University Press, Cambridge.

Montévil, M. (2019). Measurement in biology is methodized by theory. *Biology & Philosophy*, 34(3), 35.

Montévil, M., Mossio, M., Pocheville, A., and Longo, G. (2016). Theoretical principles for biology: Variation. *Progress in Biophysics and Molecular Biology*, 122, 36–50.

Mouillot, D., Bellwood, D.R., Baraloto, C., Chave, J., Galzin, R., Harmelin-Vivien, M., Kulbicki, M., Lavergne, S., Lavorel, S., Mouquet, N., Paine, C.E.T., Renaud, J., and Thuiller, W. (2013). Rare species support vulnerable functions in high-diversity ecosystems. *PLOS Biology*, 11, e1001569.

Nicolas, V., Martínez-Vargas, J., and Hugot, J.P. (2017). *Talpa aquitania* sp. nov. (Talpidae, Soricomorpha), a new mole species from SW France and N Spain. *Mammalia*, 81(6), 641–642.

Pellens, R. and Grandcolas, P. (eds) (2016). *Biodiversity Conservation and Phylogenetic Systematics: Preserving Our Evolutionary Heritage in An Extinction Crisis*. Springer International Publishing, New York.

Troudet, J., Grandcolas, P., Blin, A., Vignes-Lebbe, R., and Legendre, F. (2017). Taxonomic bias in biodiversity data and societal preferences. *Scientific Reports*, 7, 9132.

de Vargas C., Audic S., Henry N. *et al.* (2015). Eukaryotic plankton diversity in the sunlit ocean. *Science*, 348, 1261605.

Vieites, D.R., Wollenberg, K.C., Andreone, F., Kohler, J., Glaw, F., and Vences, M. (2009). Vast underestimation of Madagascar's biodiversity evidenced by an integrative amphibian inventory. *Proceedings of the National Academy of Sciences of the USA*, 106, 8267–8272.

Wilson, E.O. (ed.) (1988). *Biodiversity*. National Academies Press, Washington.

1

Symmetry of Shapes in Biology: from D'Arcy Thompson to Morphometrics

1.1. Introduction

Any attentive observer of the morphological diversity of the living world quickly becomes convinced of the omnipresence of its multiple symmetries. From unicellular to multicellular organisms, most organic forms present an anatomical or morphological organization that often reflects, with remarkable precision, the expression of geometric principles of symmetry. The bilateral symmetry of lepidopteran wings, the rotational symmetry of starfish and flower corollas, the spiral symmetry of nautilus shells and goat horns, and the translational symmetry of myriapod segments are all eloquent examples (Figure 1.1).

Although the harmony that emanates from the symmetry of organic forms has inspired many artists, it has also fascinated generations of biologists wondering about the regulatory principles governing the development of these forms. This is the case for D'Arcy Thompson (1860–1948), for whom the organic expression of symmetries supported his vision of the role of physical forces and mathematical principles in the processes of morphogenesis and growth. D'Arcy Thompson's work also foreshadowed the emergence of a science of forms (Gould 1971), one facet of which is a new branch of biometrics, morphometrics, which focuses on the quantitative description of shapes and the statistical analysis of their variations. Over the past two decades, morphometrics has developed a methodological

---

Chapter written by Sylvain GERBER and Yoland SAVRIAMA.

framework for the analysis of symmetry. The study of symmetry is today at the heart of several research programs as an object of study in its own right, or as a property allowing developmental or evolutionary inferences. This chapter describes the morphometric characterization of symmetry and illustrates its applications in biology.

Figure 1.1. *Diversity of symmetric patterns in the living world. For a color version of this figure, see www.iste.co.uk/grandcolas/systematics.zip*

COMMENTARY ON FIGURE 1.1. – *a) Centipede (C. Brena); b) corals (D. Caron); c) orchid (flowerweb); d) spiral aloe (J. Cripps); e) nautilus (CC BY 3.0); f) plumeria (D. Finney); g) Ulysses butterfly (K. Wothe/Minden Pictures); h) capri (P. Robles/Minden Pictures); i) Angraecum distichum (E. hunt); j) desmidiale (W. Van Egmond); k) diatom (S. Gschmeissner); l) sea turtle (T. Shultz); m) radiolarian (CC BY 3.0); n) starfish (P. Shaffner).*

1.2. D'Arcy Thompson, symmetry and morphometrics

A century has passed since the original publication of D'Arcy Thompson's major work, *On Growth and Form* (1917). On the occasion of this centenary, several articles have celebrated his magnum opus, highlighting the originality and enduring influence of some of Thompson's ideas in disciplines such as mathematical biology, physical biology, developmental biology and morphometrics (see Briscoe and Kicheva (2017),

Keller (2018) and Mardia *et al.* (2018) for specific reviews of these various contributions).

Thompson's main thesis is that "physical forces", such as gravity or surface tension phenomena, play a preponderant role in the determinism of organic forms and their diversity within the living world. His structuralist conception of the diversity of forms is accompanied by a critique of the Darwinian theory of evolution, but this critique is in fact based on an erroneous interpretation of the causal context (efficient cause vs. formal cause) presiding over the emergence of forms (Medawar 1962; Gould 1971, 2002, p. 1207).

The notion of symmetry is present in the backdrop throughout the book. The successive chapters go through the different orders of magnitude of the organization of life and the physical forces that prevail at each of these scales. The skeletons of radiolarians, the spiral growth of mollusk shells and the diversity of phyllotactic modes are some of the examples illustrating the pivotal role of symmetry in the architecture of biological forms. This interest in symmetry is in line with the work of Ernst Haeckel, whose book *Kunstformen der Natur* (Haeckel 1899) offers bold anatomical representations emphasizing (and sometimes idealizing) the exuberance and sophistication of symmetry patterns found in nature. Thompson sees the harmony and regularity of symmetric forms as the geometric manifestation of the mathematical principles that establish a fundamental basis for his theory of forms.

This emphasis on geometry finds its clearest expression in the last and most famous chapter of the book, "On the theory of transformation, or the comparison of related forms" (Arthur 2006). Thompson proposes a method for comparing the forms between related taxa, based on the idea of (geometric) transformation from one form to another, by means of continuous deformations of varying degrees of complexity. The morphological differences (location and magnitude) are then graphically expressed by applying the same transformation to a Cartesian grid placed on the original form. In spite of his admiration for mathematics, Thompson's approach nevertheless remains qualitative and without a formal mathematical framework for its empirical implementation. These graphical representations will, however, have a considerable conceptual impact on biologists working on the issues of shape and shape change. Several more or less elegant and operational attempts at quantitative implementation were made during the 20th century (see Medawar (1944), Sneath (1967) and Bookstein (1978), for example), up until the formulation of deformation grids using the *thin-plate splines* technique proposed by Bookstein (1989), which is still in use today.

Beyond the questions of symmetry, Thompson's idea of combining geometry and biology to study shapes remains at the heart of the principles and methods of modern morphometrics (Bookstein 1991, 1996).

1.3. Isometries and symmetry groups

In this section, we propose an overview of the mathematical characterization of the notion of symmetry, as it can be applied to organic forms. The physical environment in which living organisms evolve is comparable to a three-dimensional Euclidean space. It is denoted by $\mathbf{E}^3$. A geometric figure is said to be *symmetric* if there are one or more transformations which, when applied to the figure, leave it unchanged. Symmetry is thus a property of invariance to certain types of transformations. These transformations are called *isometries*.

Formally, an isometry of the Euclidean space $\mathbf{E}^3$ is a transformation T: $\mathbf{E}^3 \rightarrow \mathbf{E}^3$ that preserves the Euclidean metric, that is, a transformation that preserves lengths (Coxeter 1969; Rees 2000):

$$d(T(\mathrm{x}),T(\mathrm{y})) = d(\mathrm{x},\mathrm{y})$$

for all x and y points belonging to $\mathbf{E}^3$.

The different isometries of $\mathbf{E}^3$ are obtained by combining rotation and translation ($\mathrm{x} \mapsto R\mathrm{x} + t$, where R is an orthogonal matrix of order 3 and t is a vector of $\mathbb{R}^3$). They include the identity, translations, rotations around an axis, screw rotations (rotation around an axis + translation along the same axis), reflections with respect to a plane, glide reflections (reflection with respect to a plane + translation parallel to the same plane) and rotatory reflections (rotation around an axis + reflection with respect to a plane perpendicular to the axis of rotation). The set of isometries for which an object is invariant constitutes the *symmetry group* of the object.

For biologists wishing to explore symmetry in an organism, the correct identification of the symmetry group is important because it conditions the relevance of the morphometric analysis to come. The symmetry group of the organic form is always a subgroup of the isometries of $\mathbf{E}^3$. In particular, the translation has no exact equivalent in biology, since the physical extension of an organism is finite (the finite repetition of arthropod segments, for example). Translational symmetry is therefore approximate. The other isometries form a finite subgroup of Euclidean isometries, including the cyclic and dihedral groups, as well as the tetrahedral, octahedral and icosahedral groups of the Platonic solids that were dear to Thompson (1942, Chapter 9). It appears that, essentially, the symmetry patterns of biological shapes correspond to cyclic groups (rotational symmetry of order n alone [C_n], or combined with a plane of symmetry perpendicular to the axis of rotation [C_{nh}], or with n planes of symmetry passing through the axis of rotation [C_{nv}]). The bilateral symmetry of bilaterians corresponds, for example, to the group C_{1v}, in other words, a rotation of $2\pi/1$ (identity) combined with a reflection across a plane passing through the rotation axis.

1.4. Biological asymmetries

Another aspect of the imperfect nature of biological symmetry rests on the existence of deviations from the symmetric expectation (Ludwig 1932). These deviations manifest themselves to varying degrees and have distinct developmental causes. Let us consider the general case of a biological structure, whose symmetry emerges from the coherent spatial repetition of a finite number of units (e.g. the two wings of the drosophila, the five arms of the starfish). Different types of asymmetry are recognized (see also Graham *et al.* (1993) and Palmer (1996, 2004)):

– *directional asymmetry* corresponds to the case where one of the units tends to systematically differ from the others in terms of size or shape. A classic example is the narwhal, whose "horn" is in fact the enlarged canine tooth of the left maxilla, while the vestigial right canine tooth remains embedded in the gum;

– *antisymmetry* is comparable in magnitude to directional asymmetry, but the unit that differs from the others in size or shape is not the same from one individual to another. The claws of the fiddler crab show this type of asymmetry, the most developed claw being the right or the left, depending on the individual;

– *fluctuating asymmetry* is an asymmetry of very small magnitude and is therefore much more difficult to detect. It is the result of random inaccuracies in the developmental processes during the formation of the units that compose the biological structure. Fluctuating asymmetry is *a priori* always present, even if it is not always measurable. Its magnitude is considered a measure of developmental precision and has often been used (albeit sometimes controversially) as a marker of stress.

Geometrically, the morphological variation in a sample of biological shapes exhibiting a symmetric arrangement can thus be decomposed into symmetric and asymmetric variations. There is only one way to be perfectly symmetric with respect to the symmetry group of the considered structure (this is the case when all the isometries of the group are respected), but there are one less many ways to deviate from perfect symmetry as there are isometries in the group. Thus, the total variation always includes one symmetric component and at least one asymmetric component. Geometric morphometrics offers mathematical and statistical tools to quantify and explore this empirical variation.

1.5. Principles of geometric morphometrics

We are limiting ourselves here to the morphometric framework, in which the morphology of a biological structure is described by a configuration of p landmarks (Bookstein 1991; Rohlf and Marcus 1993; Adams *et al.* 2004). These landmarks, ideally defined on anatomical criteria, must be recognizable on all n specimens of the sample. Their digitization in $k = 2$ or 3 dimensions (depending on the geometry of the object considered) provides a description of each specimen as a series of pk coordinates. The comparison of the two or more objects thus characterized is done by superimposing their landmark configurations. The method most commonly used today is the Procrustes superimposition (Dryden and Mardia 1998). The underlying idea is that the shape of an object can be formally defined as the geometric information that persists once it is freed from differences in scale, position and orientation (Kendall 1984). These non-informative differences in relation to the shape are eliminated by scaling, translation and rotation, so as to minimize the sum of the squared distances between homologous landmarks (Figure 1.2). The residual variation provides a measure of the shape difference between the two objects, the *Procrustes distance*, which constitutes the metric of the *shape space*. In this space, each point corresponds to a shape, and two shapes are all the more similar, the smaller the Procrustes distance is between the points that represent them.

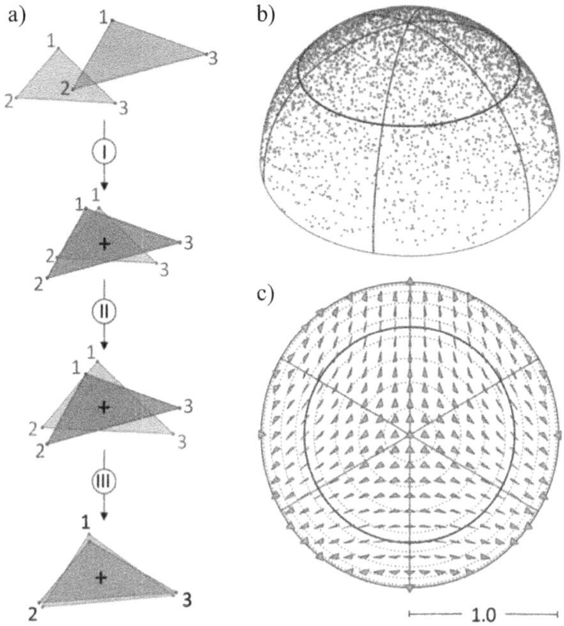

Figure 1.2. *Principles of geometric morphometrics illustrated for the simple case of triangles (three landmarks in two dimensions). For a color version of this figure, see www.iste.co.uk/grandcolas/systematics.zip*

COMMENT ON FIGURE 1.2. – *a) Two objects described by homologous configurations of landmarks are subjected to three successive transformations eliminating the differences in position (I), scale (II) and orientation (III), in order to extract a measure of their shape difference: the Procrustes distance. b) The Procrustes distance is the metric of the shape space, a non-Euclidean space in which each point corresponds to a unique shape. In applied morphometrics, researchers work with a space related to the shape space called the Procrustes (hyper)hemisphere. c) The non-linearity of this space requires projecting data onto a tangent space before testing biological hypotheses by using traditional statistical methods.*

Empirical applications of morphometrics are generally carried out in another space, the Procrustes (hyper)hemisphere, mathematically linked to the shape space, and efficiently estimating distances between shapes when the studied morphological variation is small compared to the possible theoretical variation (which is the case for biological applications). The shape space and the Procrustes hemisphere have non-Euclidean geometries and, for practical reasons (the application of standard statistical methods), the distribution of objects in either of these spaces is approximated by their distribution in a *tangent space* of the space under consideration, at the point corresponding to the mean shape of the sample (see Dryden and Mardia (1998) for more details).

1.6. The treatment of symmetry in morphometrics

Beyond their classical applications in systematics, paleontology, ecology and evolution, geometric morphometric methods have been extended over the last two decades to the analysis of the symmetry of shapes (Klingenberg and McIntyre (1998) and Mardia *et al.* (2000), and see Klingenberg (2015) for a review). As mentioned above, it is then possible to decompose the measured morphological variation into symmetric and asymmetric components, capturing the effects of distinct biological processes. When considering the symmetric organization of biological structures, two classes of symmetry are recognized by morphometricians: *matching symmetry* and *object symmetry* (Mardia *et al.* 2000; Klingenberg *et al.* 2002).

Matching symmetry describes the case where the repeated units that give the biological structure its symmetric architecture are physically disconnected from each other. This is, for example, the case for diptera wings that are attached individually to the second thoracic segment (bilateral symmetry with respect to the median plane of bilaterians), or for the arrangement of petals in many flowers (rotational symmetry). A configuration of landmarks is defined for the repeated unit (e.g. the wing) and digitized for all units (right and left wings). The configurations are then oriented in a comparable way (reflection of the left wings, so that they can be superimposed on the right wings) and analyzed using the Procrustes method (Figure 1.3(a)). In the shape space, a sample of n structures made up of m repeated units appears as m clouds of n points, whose relative overlap indicates the degree of symmetry of the structure. The more the repetition of the units is faithful to the symmetry group, the closer the different units of the same structure are and the more the clouds overlap (Figure 1.3(b)).

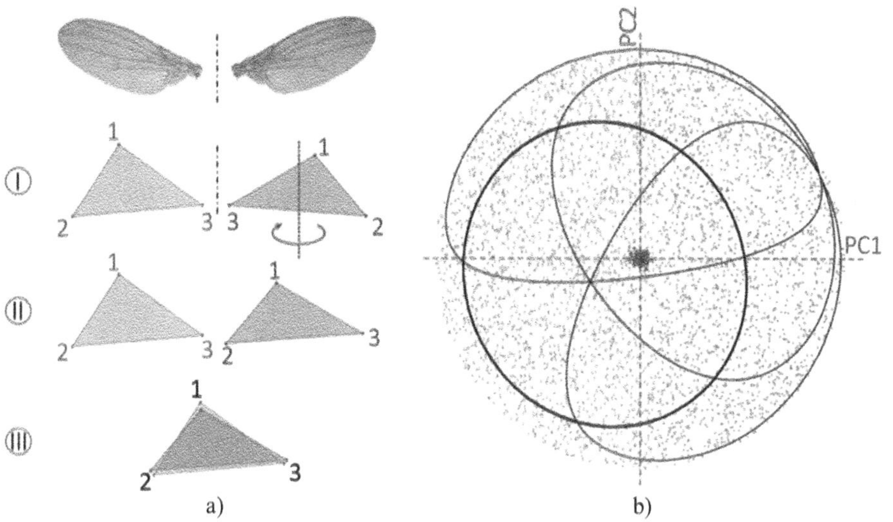

Figure 1.3. *Morphometric analysis of matching symmetry. For a color version of this figure, see www.iste.co.uk/grandcolas/systematics.zip*

COMMENT ON FIGURE 1.3. – *a) In the case of bilateral symmetry (the shape is likened to a triangle to facilitate visualization), the unit on one side (blue) is reflected (I) to match its landmark configuration with that of the same unit on the other side (green) (II). The original and reflected units of all the individuals in the sample are then superimposed by the Procrustes method. b) Each individual is then defined by two points in the tangent space corresponding to each of its repeated units. The relative position of the points and clouds of points representing the two sides (blue and green here) describes the degree of deviation from perfect bilateral symmetry at the individual and sample levels. A principal component analysis allows the visualization of the tangent space with a reduced number of dimensions.*

In the case of object symmetry, the symmetry operator is an integral part of the biological structure. The *m* repeated units are thus physically connected and their relative arrangement intervenes in the emergence of the symmetric pattern of the structure. This is, for example, the case for the right and left halves of the human face (bilateral symmetry with respect to the median plane passing through the skull) or for the arms of the starfish (rotational symmetry with respect to the antero-posterior axis passing through the center of the body). For object symmetry, a single configuration of landmarks is defined for the whole structure (and not one for each unit as in the case of matching symmetry). If q is the number of isometries of the symmetry group of the structure, then q copies of the configuration are created and transformed accordingly. All of the configurations and their copies are then superimposed (Figure 1.4(a)). In the shape space, a sample of n structures appears as q clouds of n points, where the relative overlap of the copies reflects the precision of the symmetric pattern of the structure (Figure 1.4(b)).

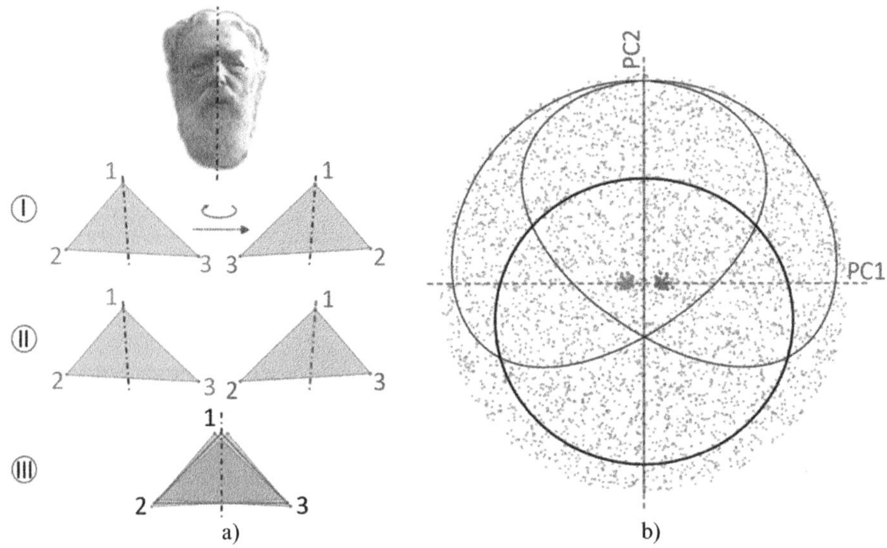

Figure 1.4. *Morphometric analysis of object symmetry. For a color version of this figure, see www.iste.co.uk/grandcolas/systematics.zip*

COMMENT ON FIGURE 1.4. – *a) In the case of bilateral symmetry (the shape is likened to a triangle for easier visualization), the original structure (orange) has a plane of symmetry. A copy of the structure is generated (the bilateral symmetry group includes two isometries: identity and reflection), reflected (I) and relabeled to ensure the correspondence between homologous landmarks (II). The original and reflected structures of all the individuals in the sample are then superimposed by the Procrustes method. b) Each individual is then defined by two points in the tangent space corresponding to each of its copies (including the original). The relative position of the points and clouds of points representing the two sides (orange and green here) describes the degree of deviation of perfect bilateral symmetry at the individual and sample levels. A principal component analysis not only allows the visualization of the tangent space with a reduced number of dimensions, but also allows the separation of the symmetric and asymmetric components.*

In both cases, the statistical exploration of the shape space allows the separation of the symmetric and asymmetric components of the shape variation. However, object symmetry has a specificity. It has been shown that the different symmetric and asymmetric components occupy complementary and orthogonal subspaces of the shape space (Kolamunnage and Kent 2003, 2005). Mathematically, the shape space is the *direct sum* of the symmetric and asymmetric subspaces. These components are easily separable and their morphological meaning can be interpreted by principal component analysis (Figure 1.5).

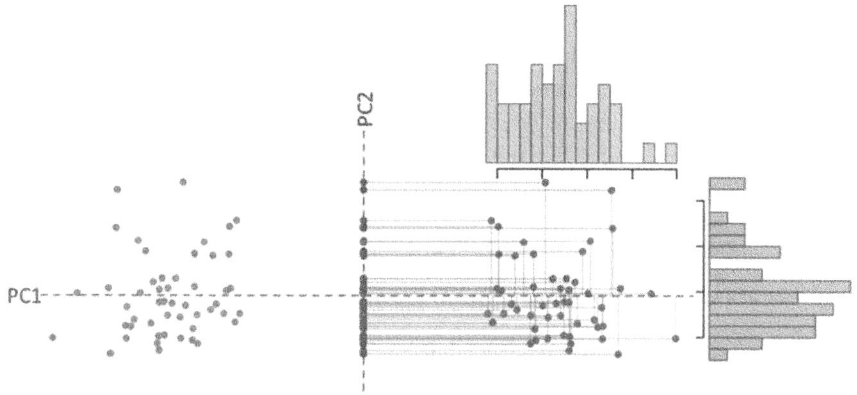

Figure 1.5. *Morphometric analysis of object symmetry and partition of the symmetric and asymmetric components. For a color version of this figure, see www.iste.co.uk/grandcolas/systematics.zip*

COMMENT ON FIGURE 1.5. – *The arrangement of the copies of the symmetric structure, generated according to its symmetry group, is such that a principal component analysis automatically separates the symmetric and asymmetric components. Here, we consider a structure similar to a triangle with a plane (axis) of symmetry. The tangent space thus has two dimensions (see Figure 1.2). We observe two clouds of points corresponding to the shapes of the original structures (orange) and to the shapes of the reflected and relabeled structures (green). The principal component analysis defines two main components, which are perfectly separated: the symmetric component, distributed along PC2 (red histogram) and describing the inter-individual variation, obtained by averaging the pairs of points corresponding to the same individual; the asymmetric component, distributed along PC1 (green histogram) and describing the intra-individual variation, obtained by calculating the difference between its original or reflected shape (orange or green points) and its symmetric shape (red points), for each individual.*

1.7. Some examples of applications

Bilateral symmetry of bilaterians was the first type of symmetry to be studied using the tools of geometric morphometrics (especially with *Drosophila* and the mouse as models). These studies are based on the methodological extension of fluctuating asymmetry analyses (Leamy 1984; Palmer and Strobeck 1986), originally based on the measurement of right–left differences of simple traits (linear measurements), to the study of shape as a highly multidimensional phenotypic trait (Klingenberg and McIntyre 1998; Mardia *et al.* 2000). Leamy *et al.* (2015), for example, using the fluctuating asymmetry in the size and shape of mouse mandibles to explore the genetic architecture of developmental instability. Quantitative trait locus (QTL) analysis underlines the epistatic genetic basis of fluctuating asymmetry, and suggests that the genes involved in the developmental stability of the mandible are the same as those controlling its shape and size.

The generalization of this morphometric framework to any type of symmetry has extended its scope to a wide variety of taxa and, in particular, to flowering plants (Savriama and Klingenberg 2011; Savriama 2018). Corolla symmetry is indeed involved in multiple aspects of the evolution of flowering plants, and morphometrics allows the statistical testing of adaptive hypotheses. For example, in *Erysimum mediohispanicum*, Gómez and Perfectti (2010) have shown the impact of the shape of the corolla (and its deviation from the expected symmetry) on the selective value of the plant: pollinators (bees, bombyliids) significantly prefer flowers with bilaterally symmetric corollas (zygomorphism).

The decomposition of the variation into symmetric and asymmetric components can also be relevant in systematics. For example, Neustupa (2013) demonstrated the possible discrimination of different species of *Micrasterias* (single-cell green algae of the Desmidiales order) according to the share of symmetric and asymmetric components, relative to the two orthogonal planes of symmetry that characterize the cell shape.

Measures of fluctuating asymmetry are also used to infer patterns of developmental integration, in other words, the modular organization of phenotypes resulting from differential interactions between developmental processes (Klingenberg 2008).

Savriama *et al.* (2016) quantified the fluctuating "translational" asymmetry in order to assess the developmental cost of segmented modular organization in eight species of soil centipedes (Geophilomorpha). The results did not show any impact of the degree of modularity (number of segments) on developmental precision, rejecting the hypothesis of a "cost" of modularity.

Finally, the architecture of some organisms or organic structures may combine several hierarchically arranged patterns of symmetry. This is, for example, the case for Aristotle's lantern, the masticatory apparatus of sea urchins, which, in regular sea urchins, combines bilateral and rotational symmetries (Savriama and Gerber 2018). The lantern exhibits the fifth-order rotational symmetry that is typical of echinoderms, and results from the repetition of a composite skeletal unit (hemipyramids + epiphyses) with bilateral symmetry. Analysis of the symmetric architecture of the lantern revealed a torsion (directional asymmetry) of the hemipyramids, contributing to the functioning of the lantern, and patterns of fluctuating asymmetry reflecting the spatialization of the skeletal precursors during the morphogenesis of the lantern.

1.8. Conclusion

The symmetry of biological forms, initially an object of curiosity and fascination, has become an important research topic in several branches of biological sciences in recent decades. The understanding of the developmental processes involved in the morphogenesis of symmetric phenotypes is a major issue in developmental and evolutionary biology (see Citerne *et al.* (2010), for example). In parallel to these genetic and developmental approaches, morphometrics has established a rigorous methodological framework for the analysis of symmetric shapes. Beyond the characterization of the symmetry of shapes, these approaches also quantify the deviations from the symmetric expectation. Their statistical analysis allows inferences to be made with regards to the architecture of complex phenotypes (genetic modularity, developmental modularity) and their variational properties (evolutionary modularity, evolvability). Coupled with molecular and developmental approaches, the recent generalization of the morphometric framework to all types of symmetry thus opens up new ways to describe, study and understand the origin and evolution of symmetries in the living world.

1.9. References

Adams, D.C., Rohlf, F.J., and Slice, D.E. (2004). Geometric morphometrics: Ten years of progress following the "revolution". *Italian Journal of Zoology*, 71, 5–16.

Arthur, W. (2006). D'Arcy Thompson and the theory of transformations. *Nature Reviews Genetics*, 7, 401–406.

Bookstein, F.L. (1978). *The Measurement of Biological Shape and Shape Change*. Springer, New York.

Bookstein, F.L. (1989). Principal warps: Thin-plate splines and the decomposition of deformation. *IEEE Transactions on Pattern Analysis and Machine Intelligence*, 11, 567–585.

Bookstein, F.L. (1991). *Morphometric Tools for Landmark Data: Geometry and Biology*. Cambridge University Press, Cambridge.

Bookstein, F.L. (1996). Biometrics, biomathematics and the morphometric synthesis. *Bulletin of Mathematical Biology*, 58, 313–365.

Briscoe, J. and Kicheva, A. (2017). The physics of development 100 years after D'Arcy Thompson's "On Growth and Form". *Mechanisms of Development*, 145, 26–31.

Chaplain, M.A.J., Singh, G.D., and McLachlan, J.C. (1999). *On Growth and Form: Spatio-temporal Pattern Formation in Biology*. John Wiley & Sons, New York.

Citerne, H., Jabbour, F., Nadot, S., and Damerval, C. (2010). The evolution of floral symmetry. *Advances in Botanical Research*, 54, 85–137.

Coxeter, H.S.M. (1969). *Introduction to Geometry*. John Wiley & Sons, New York.

Dryden, I.L. and Mardia, K.V. (1998). *Statistical Shape Analysis*. John Wiley & Sons, New York.

Gómez, J.M. and Perfectti, F. (2010). Evolution of complex traits: The case of *Erysimum* corolla shape. *International Journal of Plant Sciences*, 171, 987–998.

Gould, S.J. (1971). D'Arcy Thompson and the science of form. *New Literary History*, 2, 229–258.

Gould, S.J. (2002). *The Structure of Evolutionary Theory*. Harvard University Press, Cambridge.

Graham, J.H., Freeman, D.C., and Emlen, J.M. (1993). Antisymmetry, directional asymmetry, and dynamic morphogenesis. *Genetica*, 89, 121–137.

Haeckel, E. (1904). *Kunstformen der Natur*. Bibliographischen Instituts, Leipzig.

Keller, E.F. (2018). Physics in biology – Has D'Arcy Thompson been vindicated? *The Mathematical Intelligencer*, 40, 33–38.

Kendall, D.G. (1984). Shape manifolds, procrustean metrics, and complex projective spaces. *Bulletin of the London Mathematical Society*, 16, 81–121.

Klingenberg, C.P. (2008). Morphological integration and developmental modularity. *Annual Review of Ecology, Evolution, and Systematics*, 39, 115–132.

Klingenberg, C.P. (2015). Analyzing fluctuating asymmetry with geometric morphometrics: Concepts, methods, and applications. *Symmetry*, 7, 843–934.

Klingenberg, C.P. and McIntyre, G.S. (1998). Geometric morphometrics of developmental instability: Analyzing patterns of fluctuating asymmetry with Procrustes methods. *Evolution*, 52, 1363–1375.

Klingenberg, C.P., Barluenga, M., and Meyer, A. (2002). Shape analysis of symmetric structures: Quantifying variation among individuals and asymmetry. *Evolution*, 56, 1909–1920.

Kolamunnage, R. and Kent, J.T. (2003). Principal component analysis for shape variation about an underlying symmetric shape. In *Stochastic Geometry, Biological Structure and Images*, Aykroyd, R.G., Mardia, K.V., and Langdon, M.J. (eds). Leeds University Press, Leeds, 137–139.

Kolamunnage, R. and Kent, J.T. (2005). Decomposing departures from bilateral symmetry. In *Quantitative Biology, Shape Analysis, and Wavelets*, Barber, S., Baxter, P.D., Mardia, K.V., and Walls, R.E. (eds). Leeds University Press, Leeds, 75–78.

Leamy, L. (1984). Morphometric studies in inbred and hybrid house mice. V. Directional and fluctuating asymmetry. *The American Naturalist*, 123, 579–593.

Leamy, L., Klingenberg, C., Sherratt, E., Wolf, J., and Cheverud, J. (2015). The genetic architecture of fluctuating asymmetry of mandible size and shape in a population of mice: Another look. *Symmetry*, 7, 146–163.

Ludwig, W. (1932). *Das Rechts-Links Problem im Tierreich und beim Menschen*. Springer, Berlin.

Mardia, K.V., Bookstein, F.L., and Moreton, I.J. (2000). Statistical assessment of bilateral symmetry of shapes. *Biometrika*, 285–300.

Mardia, K.V., Bookstein, F.L., Khmabay, B.S., and Kent, J.T. (2018). Deformations and smile: 100 years of D'Arcy Thompson's "On Growth and Form". *Significance*, 15, 20–25.

Medawar, P.B. (1944). The shape of a human being as a function of time. *Proceedings of the Royal Society, Series B*, 132, 133–141.

Medawar, P.B. (1962). D'Arcy Thompson and growth and form. *Perspectives in Biology and Medicine*, 220–232.

Neustupa, J. (2013). Patterns of symmetric and asymmetric morphological variation in unicellular green microalgae of the genus *Micrasterias* (Desmidiales, Viridiplantae). *Fottea*, 13, 53–63.

Palmer, A.R. (1996). Waltzing with asymmetry. *Bioscience*, 46, 518–532.

Palmer, A.R. (2004). Symmetry breaking and the evolution of development. *Science*, 306, 828–833.

Palmer, A.R. and Strobeck, C. (1986). Fluctuating asymmetry: Measurement, analysis, patterns. *Annual Review of Ecology and Systematics*, 17, 391–421.

Rees, E.G. (2000). *Notes on Geometry*. Springer, Berlin.

Rohlf, F.J., Marcus, L.F. (1993). A revolution in morphometrics. *Trends in Ecology and Evolution*, 8, 129–132.

Savriama, Y. (2018). A step-by-step guide for geometric morphometrics of floral symmetry. *Frontiers in Plant Science*, 9, 1433.

Savriama, Y. and Gerber, S. (2018). Geometric morphometrics of nested symmetries unravels hierarchical inter-and intra-individual variation in biological shapes. *Scientific Reports*, 8, 18055.

Savriama, Y. and Klingenberg, C.P. (2011). Beyond bilateral symmetry: Geometric morphometric methods for any type of symmetry. *BMC Evolutionary Biology*, 11, 280.

Savriama, Y., Vitulo, M., Gerber, S., Debat, V., and Fusco, G. (2016). Modularity and developmental stability in segmented animals: Variation in translational asymmetry in geophilomorph centipedes. *Development Genes and Evolution*, 226, 187–196.

Sneath, P.H.A. (1967). Trend surface analysis of transformation grids. *Journal of Zoology*, 151, 65–122.

Thompson, D.W. (1917). *On Growth and Form*. Cambridge University Press, Cambridge.

2

Impact of a Point Mutation in a Protein Structure

2.1. Composition

Proteins are involved in most cellular functions at all levels, from DNA duplication to chemical metabolism, cell structuring and signal transmission. Despite their very varied activities, these molecules are quite homogeneous: proteins[1] are polymers composed of 20 base units, the amino acids (or residues when polymerized). The latter are composed of a central carbon atom (Cα) linked to an amino group (NH_2), a carboxylic group (COOH), a hydrogen atom and one of 20 different chemical groups called "side chains" (Figure 2.1). Residues and their different chemical functions allow the functional diversity of proteins. In the polypeptide chain, the α-carboxylic group of an amino acid is linked to the α-amino group of the next amino acid through an amide bond (peptide bond –CO–NH–, Figure 2.2). Most natural proteins contain between 50 and 2,000 amino acid residues. The unbranched chain of residues is oriented: it starts at the amino end (N-terminal) and ends at the carboxy end (C-terminal). The chain of atoms regularly repeating the peptide bonds is called the "peptide backbone". The peptide bond is rigid and flat because of the partial double bond character of the –CO–NH– bond, but rotations are possible around the other bonds of the backbone.

Chapter written by Mathilde CARPENTIER and Jacques CHOMILIER.

1 Or peptides, which are chains containing less than 50 amino acids linked by peptide bonds.

Proteins can be roughly divided into four classes according to their morphology: globular proteins, which are in an aqueous environment, fibrous proteins, which form large aggregates and mostly constitute the cytoskeleton, membrane proteins and so-called "disordered" proteins, which are generally small and have no inherent fixed structure. Here, we will only discuss the first category of proteins, which are the globulars.

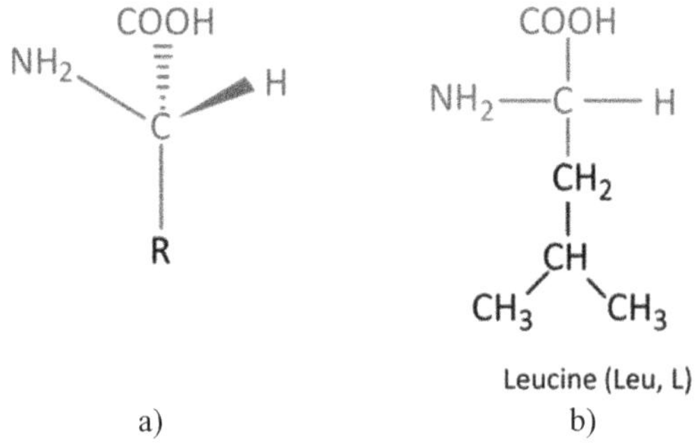

Figure 2.1. *a) General structure of an amino acid; b) chemical formula of leucine*

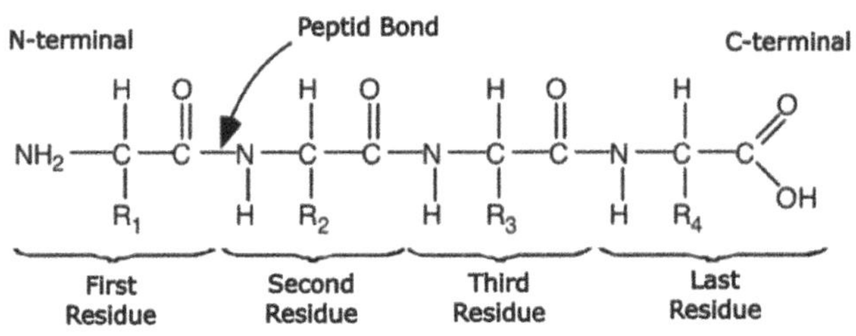

Figure 2.2. *A polypeptide consisting of four amino acids (tetrapeptide)*

2.2. Folding

Proteins adopt a specific spatial organization, most often called a "structure". This structure is crucial for their function. This relationship between structure and function, established by Emil Fisher at the end of the 19th century, is the foundation of structural biology. Methods used for determining the structure of proteins have evolved considerably, but the method of choice remains as X-ray diffraction, which requires the protein to be crystallized. The data bank listing protein structures (as well as nucleic acids and some sugars) is the PDB (Protein Data Bank) (Berman *et al.* 2000). The number of resolved protein structures is growing rapidly from year to year. A protein could theoretically adopt a large number of three-dimensional conformations, but most of them spontaneously fold into a particular and unique stable form. This particular shape is due to the fact that the peptide backbone groups and side chains interact with each other and with water. Thus, some conformations have more stabilizing interactions than others and are therefore favored (Alberts *et al.* 1994). The paradigm of the relationship between the protein sequence and its three-dimensional (3D) structure comes from Christian Anfinsen's studies on ribonuclease (Anfinsen 1973). Anfinsen showed that proteins isolated in solution can regain their original active conformation after denaturation. Therefore, the conclusion was that all the information needed to fold a protein must be inherent to its amino acid order (Alberts *et al.* 1994). Other studies have also drawn the same conclusions, leading to the general theory that the amino acid sequence of a protein specifies its conformation (Stryer 1994).

"Water-soluble" proteins fold into a compact globular form (unlike fibrous, membrane and "disordered" proteins). The hydrophobic nature of certain amino acids makes this compact folding necessary. Indeed, the side chains of the non-polar residues are hydrophobic and are grouped together within the globular structure of the protein – isolated from the surrounding water – while the polar residues, and the polar groups of the backbone, are hydrophilic and form hydrogen bonds with water or with each other. This spatial distribution of hydrophobicity has been known since the middle of the 20th century, in particular thanks to the work of Bressler and Talmud (1944), who described a globular protein as a micelle, with a predominantly hydrophobic core surrounded by a predominantly hydrophilic crown. The hydrophobic and hydrogen bonds are largely responsible for the stability of the protein structure, obtained under the effect of the only available energy source, thermal agitation. These continuous fluctuations lead to more or less rapid displacements around stable local conformations, called secondary structures. *Alpha* helices (30% of the residues) and *beta* strands (20% of the residues) are the most frequent. These local structures are stabilized by hydrogen bonds due to the particular values of the dihedral angles of the peptide backbone.

2.3. Substitution(s) in protein structures

The substitution of one amino acid by another (a mutation) is one of the fundamental events of molecular evolution, with variable consequences in proteins. The majority of mutations have no major effect on the phenotype – they are neutral – but some of them can cause disease (Studer *et al.* 2013). Indeed, local changes may occur in binding sites with other molecules and can thus affect the function of proteins (Gong *et al.* 2009), but long-term effects on the overall structure can also be observed (Zhou *et al.* 2007).

Many studies agree that the majority of substitutions have no significant effect on the overall structure, stability or function of the protein. As a matter of fact, it has been shown that 75% of the amino acids can be modified without significant alteration of the protein structure (Sander and Schneider 1991; Shakhnovich and Gutin 1991; Schaefer and Rost 2012). These observations support the neutral hypothesis of point mutations, but it is important to keep in mind that this does not mean that all mutations are neutral: the majority of point mutations are effectively counter-selected because their impact is negative for the cell. For example, the probability that a human DNA repair enzyme, 3-methyladenine DNA glycosylase, becomes non-functional after a random mutation is 34% (± 6%), and this proportion can be extended to other families (Guo *et al.* 2004).

2.4. Effect on overall structure and function

From a structural point of view, the effect of an amino acid substitution is difficult to model because the effects can be drastic. A striking example is the L16A mutation (leucine 16 replaced by alanine) of a DNA-binding protein of *Drosophila melanogaster*, located in the homeodomain, which profoundly modifies the structure of the native protein while maintaining the three helices (Religa *et al.* 2005) (Figure 2.3).

However, the link between the structural change and functionality, or disease is not obvious or systematic. For example, in the case of human lysozymes with many known structures, two natural mutants, D67H and I56T, form amyloid fibrils in the extracellular space of multiple organs and tissues, resulting in non-neuropathic systemic amyloidosis. In the case of the D67H mutation (Figure 2.4(a)), the structure of the lysozyme is highly disrupted in two of its loops, while the structure of the I56T mutant (Figure 2.4(b)) appears little disrupted, as shown by the structures of the mutants superimposed on that of the native lysozyme. Nevertheless, without attempting to predict whether a mutation will be pathogenic or not, the modeling of the effect of a mutation on a protein has been proposed by assessing the variation in its stability.

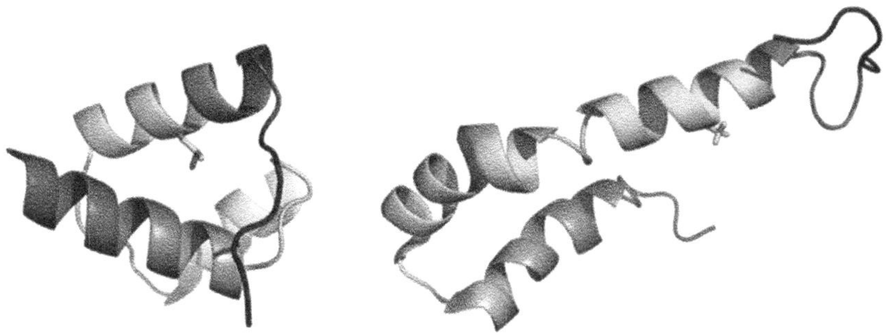

a) b)

Figure 2.3. *a) DNA binding protein of* Drosophila melanogaster *(PDB code 1enh); b) L16A mutant of the same protein (PDB code 1ztr). The mutated position is shown with its side chain. For a color version of this figure, see www.iste.co.uk/grandcolas/systematics.zip*

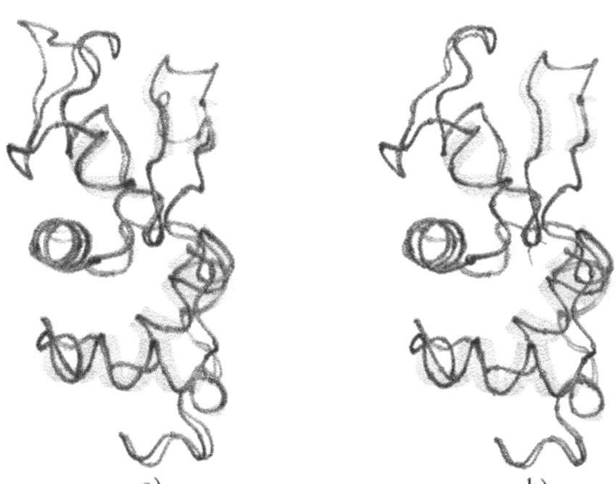

a) b)

Figure 2.4. *a) Superimposed structures of the D67H mutant (green, PDB code 1lyy) and the native structure (blue, PDB code 2nwd) of the human lysozyme; b) superimposed structures of the I56T mutant (orange, PDB code 1loz) and the native structure (blue, PDB code 2nwd) of the human lysozyme. The mutated positions are represented with their side chains. For a color version of this figure, see www.iste.co.uk/grandcolas/systematics.zip*

2.5. Effect on stability

Proteins are said to be "marginally stable": typically, there is a difference of 3–7 kcal/mol in free folding energy (ΔG) between folded and unfolded conformations. Amino acid side chain substitutions thus have a significant effect on protein stability: the effect of a single mutation is on average –0.95 kcal/mol according to the Protherm database (Gromiha and Sarai 2010), which in 2017 included the measurements from ΔΔG for 1,866 proteins with their structure. ΔΔG is the difference between the free folding energy of the native and the mutated protein.

This low protein stability is assumed to be either the result of a balance between function and stability (DePristo *et al.* 2005), or the result of a balance between destabilizing mutations and highly unstable proteins (Taverna and Goldstein 2002; Bloom *et al.* 2007; Zeldovich *et al.* 2007).

Two principles must be kept in mind:

– function is often dependent on a dynamic effect of structure;

– any protein must be degradable at a cost that is not restrictive for the cell.

Many algorithms and web servers have been developed to provide an estimate of the variation of Gibbs' free energy (ΔΔG) under the effect of a point mutation: FoldX (Guerois *et al.* 2002) and Rosetta (Kellogg *et al.* 2011) are among the best known. SPROUTS (Lonquety *et al.* 2009) is a web server combining the results of several methods. These methods try to predict whether a given mutation will be destabilizing, neutral or stabilizing. The comparison between predicted and experimental energy variations gives globally satisfactory results (Figure 2.5(a)): the correlation between the predictions of FoldX and the measurements of Protherm is 0.59. Nevertheless, the difference can be quite large in some cases (Lonquety *et al.* 2009). It is interesting to note that the prediction results are quite different when comparing wild to mutant and mutant to wild, in absolute value (Figure 2.5(b)).

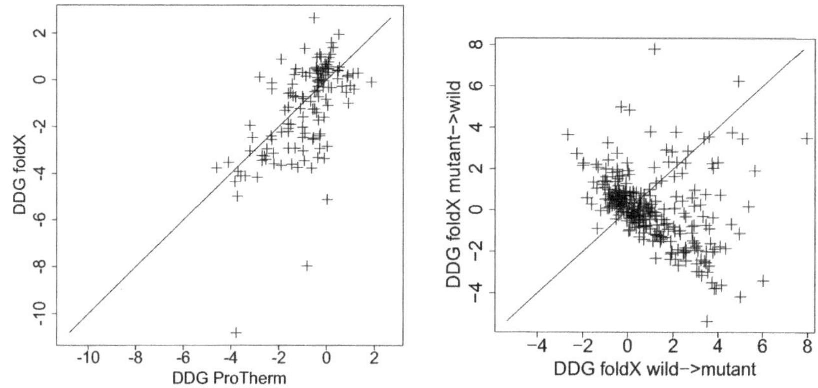

a) b)

Figure 2.5. *Comparison of predicted ΔΔG by FoldX and those experimentally measured (Protherm)*

COMMENT ON FIGURE 2.5. – *a) Prediction by FoldX of ΔΔG for 130 proteins present in Protherm and belonging to the 11 families in which at least 20 different point mutant structures are known (see section 2.6 for a description of this dataset). b) Prediction by FoldX of ΔΔG for the families in which at least 20 different point mutant structures are known. The abscissa shows the predicted value of ΔΔG from the native to the mutant, and the ordinate shows the predicted value of ΔΔG from the mutant to the native.*

2.6. Effect on the peptide backbone

In the calculation methods of ΔΔG mentioned above, the main chain is fixed; only the side chains are allowed to move. The range of motion of the side chain is extracted from the rotamer libraries (Dunbrack 2002), and their degrees of freedom strongly depend on the conformation of the main chain. Taking the deformation of the main chain into account should improve our understanding of the consequences of substitution. Nevertheless, it is legitimate to ask whether the backbone is really disrupted by a simple substitution, at least in the one we can observe, and whether this disruption is measurable with current techniques. Nowadays, the resolution of the structure of proteins determined by crystallography and X-ray diffraction is of the order 1 Å, which means that the precision of the structures obtained is of the order 1/10 Å (Luzzati 1952). The displacement of the backbone atoms caused by substitution is typically also 1/10 Å, which puts us at the limit of resolution.

In addition, protein structures of identical sequences can vary greatly due to different protein–protein interactions, interactions with different ligands or solvents (Kosloff and Kolodny 2008). However, provided that a sufficient number of structures are available, the effect of the mutation can be distinguished from "noise", in other words, from statistical fluctuations due to other sources, such as exposure to solvents or belonging to a secondary structure (Shanthirabalan *et al.* 2018). This requires measuring the variations between the native and mutated proteins, and considering both the global variability and local flexibility of the structure.

The objective is therefore to measure whether a given structure (generally the mutant structure) is different at certain places (locally) – for example, at the place of mutation – from another so-called "reference" structure (generally the native structure). These two proteins have extremely close sequences (for a punctual mutation), so it is easy to find the alignment (correspondence) between the amino acids of the two proteins. Then, the two structures (in green and red, Figure 2.6(a)) are superimposed to minimize the distances between the corresponding pairs of $C\alpha$. We will then measure the distances between the $C\alpha$s of the two structures for small fragments (in blue, Figure 2.6(a)) of three consecutive $C\alpha$s by calculating the RMSD (Root-Mean-Square Deviation). A length of three residuals was chosen because it is the length for which the measured local effect is the strongest (Shanthirabalan *et al.* 2018). This calculation is performed for all residues and we obtain a "profile" of the RMSDs (Figure 2.6(b), graph) for a given protein pair.

In order to determine whether RMSDs tend to be larger in terms of mutations, it is necessary to be able to compare several structures that only differ from each other by a single point mutation. In PDB, there are 11 families with at least 20 mutants with a single substitution from a reference structure, for a total of 580 mutants and 11 reference structures (Shanthirabalan *et al.* 2018). When comparing the profiles between proteins of the same family, it can be observed that the regions with the largest RMSDs are often the same, regardless of the position of the mutation. Overall, some proteins "move" very little, while others "move" a lot (Figure 2.7).

It is therefore difficult to know whether a large RMSD is due to the mutation, to the intrinsic flexibility of the protein or to an overall variability resulting from the various crystallization conditions. Thus, if a comparison is made between the distribution of RMSDs centered on mutated residues and the distribution of RMSDs for all positions of the structures of the 11 families, one will observe that the two distributions significantly overlap (Figure 2.8(a)). To better differentiate the effect due to mutations, the two other sources of variability must be neutralized.

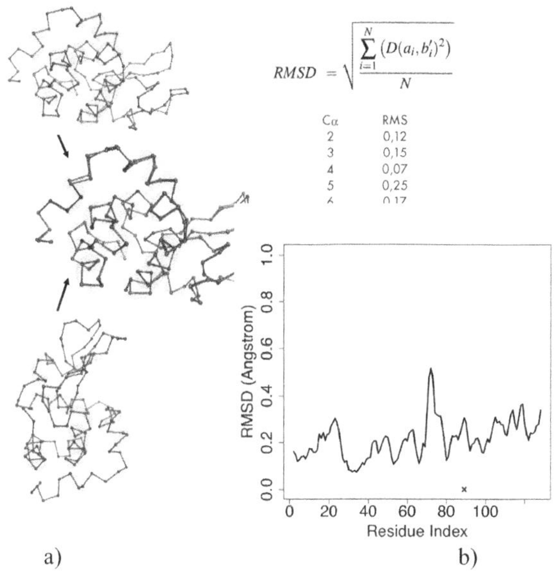

Figure 2.6. *Procedure for calculating "local" RMSDs. For a color version of this figure, see www.iste.co.uk/grandcolas/systematics.zip*

COMMENT ON FIGURE 2.6. – *a) Superimposition of two lysozyme structures. b) Their difference is measured by the RMSD calculated for the fragments of three successive residues, such as the fragment in blue. The RMSD is the root of the sum of the N distances D between the alpha carbon pairs a_i and b'_i (i.e. b_i after superimposition) divided by the N number of Cα pairs (three in our case). This calculation is performed on the whole protein. There are as many RMSDs as residues in the protein (except for the two Cαs at the extremities); a profile is obtained (graph below). The mutation is localized at the location of the cross (protein 2hef chain A, mutation I89A).*

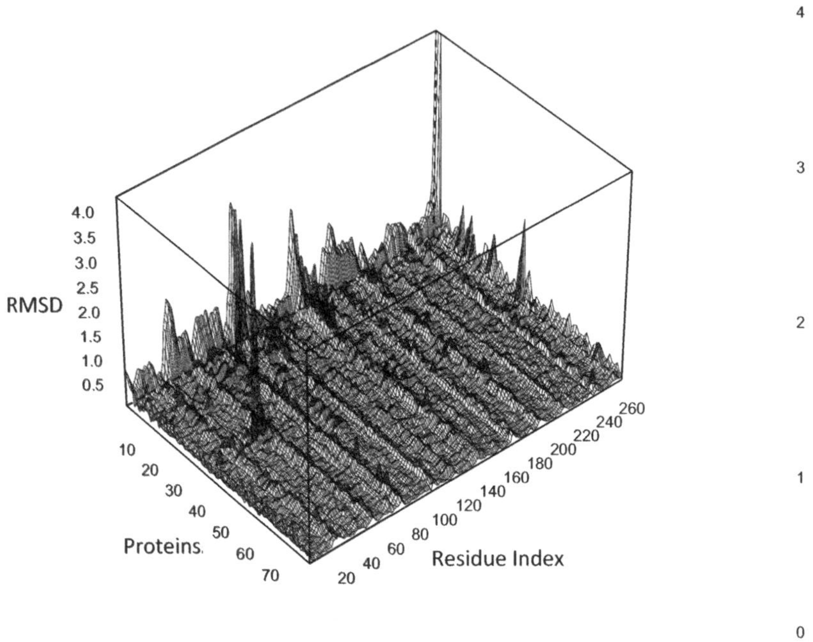

Figure 2.7. *RMSD calculated for 78 mutants of a transferase, the reference being that of* Pyrococcus horikoshii *(PDB code 2dek chain A). For a color version of this figure, see www.iste.co.uk/grandcolas/systematics.zip*

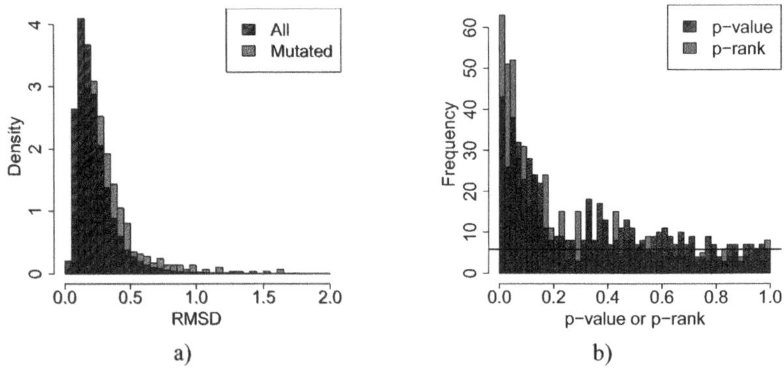

Figure 2.8. *Distribution of RMSD, p-values and p-ranks. For a color version of this figure, see www.iste.co.uk/grandcolas/systematics.zip*

COMMENT ON FIGURE 2.8. – *a) Distribution of RMSDs for all positions (crosshatched) or mutated residues alone (blue) for 580 mutations distributed in 11 families. b) Distribution of empirical p-values (crosshatched) and p-ranks (red) of mutated residues. The black line represents the uniform distribution of p-values or p-rank, in other words, the distribution expected if 580 RMSDs are randomly drawn from the crosshatched distribution of (a).*

In order to take into account the global variability due to variations in experimental conditions, the gross RMSD should not be used, but a transformation of it. Considering ranks instead of values is a robust transformation used in many statistical tests. The RMSDs in each profile are first ranked in ascending order, and then the ranks are divided by the number of RMSDs in the profile (in other words, the length of the chain). The result is dimensionless, stacked values that allow the characterization of each protein in the family. If the mutations had no particular effect on the RMSD, the distribution of these p-values should be uniform, which is not what is observed (Figure 2.8(b)). This first transformation allows the experimental variability, but not the intrinsic flexibility of the molecule, to be taken into account. Indeed, in very flexible regions, seeing as the RMSD is large, the first ranking, and thus the empirical p-value, will also always be large. It is therefore necessary to make a second classification, that of the empirical p-value, for each position in each family. The new empirical p-value is then called the p-rank in order to differentiate it from the first one.

The place where the mutation takes place is the one most likely to be disturbed, at least in intensity (with a high RMSD value). Among all the calculated RMSDs, there are 580 positions corresponding to a mutation. Also, the p-rank distribution of RMSDs centered on mutated residues is not uniform (Figure 2.8(b)). Among the top 5% of the largest RMSDs, 12% are mutation-centered; among the top 5% of empirical p-ranks, 15% are mutation-centered; and among the top 5% of p-ranks, 25% are mutation-centered. These two transformations thus allow better isolation of locally disrupted regions.

The RMSDs calculated in the least flexible regions, buried, or located in the alpha helices and beta sheets are small and have lower variability (0.21 Å ± 0.19, 0.22 Å ± 0.19, 0.19 Å ± 0.14, respectively) compared to all RMSDs (0.24 Å ± 0.25). Local disruptions of the main chain in these regions are thus more difficult to isolate, but as we take into account the flexibility of the proteins in our transformations, we can extract disruptions in these regions that, although small, are significant because the variability is low. For example, 17% of the residues are in beta strands in the whole sample. Among the segments with the largest 5% RMSD, the proportion of residues located in beta strands falls to 9%, while for the smallest 5% p-rank, it rises to 17%. We have thus effectively eliminated the bias due to the intrinsic rigidity of the beta sheets.

2.7. Conclusion

The methodology presented allows the identification of disturbed regions in protein structures by taking into account biases due to experimental variations and protein flexibility. Now that we know that mutations do indeed disrupt the main chain and that these disruptions are measurable with current techniques, it would be interesting to model them, especially to improve the predictions of ΔΔG, for which the carbon chain is rigid.

Two models exist for the accommodation of the main chain under the effect of amino acid substitution. The first (Davis *et al.* 2006) is derived from the observation of alternative atomic positions in ultra-high resolution crystallographic structures. It has been successfully used to improve Rosetta's calculation of ΔΔG (Lauck *et al.* 2010). The second (Bordner and Abagyan 2004) was constructed from data collected on 2,141 pairs of protein structures, only differing by a single point mutation. This model also improved Gibbs' prediction of free energy after a mutation. The selection method presented allows the identification of fragments where the main chain was more disrupted than expected. Using this database instead of the previous ones should improve the models.

2.8. References

Alberts, B., Bray, D., Lewis, J., Raff, M., Toberts, K., and Watson, J. (1994). *Molecular Biology of the Cell*. Garland Publishing, New York.

Anfinsen, C.B. (1973). Principles that govern the folding of protein chains. *Science*, 181, 223–230.

Berman, H.M., Westbrook, J., Feng, Z., Gilliland, G., Bhat, T.N., Weissig, H., Shindyalov, I.N., and Bourne, P.E. (2000). The Protein Data Bank. *Nucleic Acids Research*, 28, 235–242.

Bloom, J.D., Raval, A., and Wilke, C.O. (2007). Thermodynamics of neutral protein evolution [Online]. *Genetics*, 175, 255–266. Available: https://doi.org/10.1534/genetics.106.061754.

Bordner, A.J. and Abagyan, R.A. (2004). Large-scale prediction of protein geometry and stability changes for arbitrary single point mutations [Online]. *Proteins*, 57, 400–413. Available: https://doi.org/10.1002/prot.20185.

Bressler, S. and Talmud, D. (1944). On the nature of globular proteins. *Comptes rendus de l'Académie des sciences de l'URSS*, 43, 310–314.

Davis, I.W., Arendall, W.B., Richardson, D.C., and Richardson, J.S. (2006). The backrub motion: How protein backbone shrugs when a sidechain dances [Online]. *Structure*, 14, 265–274. Available: https://doi.org/10.1016/j.str.2005.10.007.

DePristo, M.A., Weinreich, D.M., and Hartl, D.L. (2005). Missense meanderings in sequence space: A biophysical view of protein evolution [Online]. *Nature Reviews Genetics*, 6, 678–687. Available: https://doi.org/10.1038/nrg1672.

Dunbrack, R.L. (2002). Rotamer libraries in the 21st century [Online]. *Current Opinion in Structural Biology*, 12, 431–440. Available: https://doi.org/10.1016/S0959-440X(02)00344-5.

Gong, S., Worth, C.L., Bickerton, G.R.J., Lee, S., Tanramluk, D., and Blundell, T.L. (2009). Structural and functional restraints in the evolution of protein families and superfamilies [Online]. *Biochemical Society Transactions*, 37, 727–733. Available: https://doi.org/10.1042/BST0370727.

Gromiha, M.M. and Sarai, A. (2010). Thermodynamic database for proteins: Features and applications [Online]. *Methods in Molecular Biology*, 609, 97–112. Available: https://doi.org/10.1007/978-1-60327-241-4_6.

Guerois, R., Nielsen, J.E., and Serrano, L. (2002). Predicting changes in the stability of proteins and protein complexes: A study of more than 1000 mutations [Online]. *Journal of Molecular Biology*, 320, 369–387. Available: https://doi.org/10.1016/S0022-2836(02)00442-4.

Guo, H.H., Choe, J., and Loeb, L.A. (2004). Protein tolerance to random amino acid change [Online]. *Proceedings of the National Academy of Sciences*, 101, 9205–9210. Available: https://doi.org/10.1073/pnas.0403255101.

Kellogg, E.H., Leaver-Fay, A., and Baker, D. (2011). Role of conformational sampling in computing mutation-induced changes in protein structure and stability [Online]. *Proteins*, 79, 830–838. Available: https://doi.org/10.1002/prot.22921.

Kosloff, M. and Kolodny, R. (2008). Sequence-similar, structure-dissimilar protein pairs in the PDB [Online]. *Proteins*, 71, 891–902. Available: https://doi.org/10.1002/prot.21770.

Lauck, F., Smith, C.A., Friedland, G.F., Humphris, E.L., and Kortemme, T. (2010). RosettaBackrub – A web server for flexible backbone protein structure modeling and design [Online]. *Nucleic Acids Research*, 38, W569–W575. Available: https://doi.org/10.1093/nar/gkq369.

Lonquety, M., Lacroix, Z., Papandreou, N., and Chomilier, J. (2009). SPROUTS: A database for the evaluation of protein stability upon point mutation [Online]. *Nucleic Acids Research*, 37, D374-9. Available: https://doi.org/10.1093/nar/gkn704.

Luzzati, V. (1952). Traitement statistique des erreurs dans la determination des structures cristallines [Online]. *Acta Crystallographica*, 5, 802–810. Available: https://doi.org/10.1107/S0365110X52002161.

Religa, T.L., Markson, J.S., Mayor, U., Freund, S.M.V., and Fersht, A.R. (2005). Solution structure of a protein denatured state and folding intermediate [Online]. *Nature*, 437, 1053–1056. Available: https://doi.org/10.1038/nature04054.

Sander, C. and Schneider, R. (1991). Database of homology-derived protein structures and the structural meaning of sequence alignment [Online]. *Proteins*, 9, 56–68. Available: https://doi.org/10.1002/prot.340090107.

Schaefer, C. and Rost, B. (2012). Predict impact of single amino acid change upon protein structure [Online]. *BMC Genomics*, 13, S4. Available: https://doi.org/10.1186/1471-2164-13-S4-S4.

Shakhnovich, E.I. and Gutin, A.M. (1991). Influence of point mutations on protein structure: Probability of a neutral mutation. *J. Theor. Biol.*, 149, 537–546.

Shanthirabalan, S., Chomilier, J., and Carpentier, M. (2018). Structural effects of point mutations in proteins [Online]. *Proteins: Structure, Function, and Bioinformatics*, 86, 853–867. Available: https://doi.org/10.1002/prot.25499.

Stryer, L. (1994). *Biochemistry*. W.H. Freeman, New York.

Studer, R.A., Dessailly, B.H., and Orengo, C.A. (2013). Residue mutations and their impact on protein structure and function: Detecting beneficial and pathogenic changes [Online]. *Biochemical Journal*, 449, 581–594. Available: https://doi.org/10.1042/BJ20121221.

Taverna, D.M. and Goldstein, D. (2002). Why are proteins marginally stable? [Online]. *Proteins*, 46(1), 105–109. Available: https://doi.org/10.1002/prot.10016.

Zeldovich, K.B., Chen, P., and Shakhnovich, E.I. (2007). Protein stability imposes limits on organism complexity and speed of molecular evolution [Online]. *Proceedings of the National Academy of Sciences*, 104, 16152–16157. Available: https://doi.org/10.1073/pnas.0705366104.

Zhou, R., Eleftheriou, M., Royyuru, A.K., and Berne, B.J. (2007). Destruction of long-range interactions by a single mutation in lysozyme [Online]. *Proceedings of the National Academy of Sciences*, 104, 5824–5829. Available: https://doi.org/10.1073/pnas.0701249104.

3

The Role of Taxonomy and Natural History in the Study of the Evolution of Eneopterinae Crickets

This chapter aims to demonstrate how *taxonomy*, *natural history* and *descriptive approaches* are paramount in the study of evolution. I first discuss taxonomy's place within comparative approaches. I then develop the example of Eneopterinae crickets by detailing the recent contributions of taxonomic work on *phylogenetic relationships*, *biogeographic patterns*, and on the exploration of the *evolution of traits* associated with *communication* in these insects.

3.1. Introduction

There are almost as many ways of understanding *biodiversity* as there are disciplines that have this ambition. This can be done at the scale of a few model organisms whose genomes are finely compared, or at the scale of whole phyla to study evolution using phylogenies. From the molecule to the ecosystem, from the individual to the phylum, it all revolves around the scale at which we wish to focus our question, and then there are the levels of observation, interpretation and integration of knowledge. There is no right or wrong positioning in this field, but rather questions that can be more or less well addressed at a given scale. These multiple points of view are complementary, as it is often necessary to adopt one in order to better understand another (Losos 2009, 2011).

Chapter written by Tony ROBILLARD.

All of this work, however, depends on a single initial scale of study: that of descriptive approaches of biodiversity, which includes disciplines such as taxonomy and nomenclature, and the description of all the traits of the species, such as their behavior and life histories. It is all of these components – known as "natural history" – that form the foundation of natural history museums, and make up their wealth of collections and knowledge.

However, the majority of research carried out until recently in biology has focused on a small number of model organisms, or failing that, on *archetypes*, which tends to minimize the work on the "rest", which constitutes the vast majority of life. In spite of a growing awareness that the understanding of life cannot be limited to the study of a few species, much remains to be done to reconcile the diversity of life with the study of the underlying biological and evolutionary mechanisms (e.g. Manger *et al.* (2008), Alfred and Baldwin (2015), Brenowitz and Zakon (2015)). Indeed, it is still all too common to consider that the accumulated knowledge of a few "representative" species of a given group is sufficient to understand its diversity, as is the case for crickets and acoustic communication (see section 3.6). However, a taxonomic group is never homogeneous, as a quick look at the phenotypes of the species around us suggests. All mammals – although highly favored in terms of model organisms (Troudet *et al.* 2017; Miralles *et al.* 2019) – are hardly comparable to mice, dogs or humans, and any fishmonger's stall has many more species than the list of teleosts that are model species (zebrafish and fugu). Does this mean that the diversity of non-model species hides a poorness of mechanisms? Or should we expect to have to reconsider many supposed truths acquired from the study of a species chosen at random, or jointly, from a huge collection of special cases?

The second aspect of the problem is the difficulty in valuing descriptive work in the current scientific and editorial landscape, which leads to complications for publication, then during research assessments, including peer review, and finally during funding applications. This difficulty is a reflection of the scientific community itself, because if the majority has little or no interest in the diversity of life, this same majority is likely to grant less intellectual or financial credit to teams wishing to do so. The direct consequence is the reinforcement, or the continuation, of the trend towards a scarcity of descriptive work.

Among the disciplines interested in the study of evolution, however, comparative approaches are particularly well considered, since evolution involves diversity. Disciplines such as taxonomy and *nomenclature* are thus in constant interaction, and even overlap with those committed to understanding the mechanisms of evolution. But even so, there is a significant tendency to minimize descriptive approaches in favor of works that are centered on questions.

The consequence is a certain disconnection between a part of the community that generates data based on natural history, from species descriptions to occurrence databases, and another part of the community that values them in a more ambitious, theoretical and statistical context. In addition, there are questions about the very legitimacy of pursuing descriptive work: why continue to describe or document species when there are already so many of which we know nothing about other than a few specimens collected during the 19th century?

In these circumstances, one way to proceed is to adopt an approach that combines descriptive approaches with studies of evolutionary questions. It is this approach that I use to study crickets of the Eneopterinae subfamily.

In this chapter, I will discuss various aspects of how taxonomy and descriptive approaches are central to the study of evolution. After a brief reminder that taxonomy is part of comparative approaches, I will present the Eneopterinae study model in detail, reviewing the recent contributions related to taxonomic work in terms of knowledge on phylogenetic relationships and biogeographic patterns. Finally, through this case study, I will address the question regarding the usefulness of taxonomic exploration to study the evolution of communication.

3.2. Taxonomy in modern comparative approaches

Beyond its role in the discovery and description of taxa, the taxonomist cultivates in-depth knowledge of a group of organisms, which is a fundamental asset in any study on the evolution and diversification of life. A mastered taxonomy makes it possible to constitute an adequate sampling in relation to a given scientific objective, taking into account the existing *classification* and its possible biases and gaps. Moreover, comparative analyses have come to realize that, in order to draw robust conclusions, it is preferable to take into account a sampling that is as exhaustive as possible of the groups under study. Without the expertise of the *taxonomist*, very few groups are likely to meet this condition. In fact, the systematics activity is very often inseparable from the taxonomic sampling itself, either through the discovery of new taxa during expeditions to collect study material, or through the revision of groups requiring taxonomic clarification (synonymous species, difficulty in assigning material to existing taxa, etc.).

Evolutionary studies are generally not limited to reconstructing phylogenetic trees. Studies consider spatiotemporal dimensions that allow the dating of diversification events and address the dynamics of distribution areas and endemism over time. Similarly, species diversification is itself an important subject of study, very often linked to phenotypic and/or ecological diversification. This work leads to syntheses that put the evolutionary history of clades and their phenotypic characteristics into perspective (see, for example, Lemey *et al.* (2009) and Baum and Smith (2012)). Such studies require multidisciplinary approaches with a central *systematic* component. In addition to naming and describing taxa, the role of the taxonomist is also to collect, compile and describe all information relating to the taxa of which he or she is a specialist. To do this, fieldwork not only contributes to taxonomy and the enrichment of character matrices for phylogenetic reconstructions, but also provides essential information on the distribution of taxa, their biology or their behavior; all this information enables the taxonomists to refine their knowledge of the group.

These comparative data cannot be limited to vague or bibliographical information; they require a considerable sampling effort (Grandcolas *et al.* 1994, 1997) because mapping an ecological or phenotypic trait on a phylogeny is not enough to understand its evolution. In most cases, the study requires upstream work that is just as important, if not more important, than the phylogenetic reconstruction itself. Distributions, behaviors, communication signals, habitats, biotopes, predator–prey relationships and competitors are all biological variables that characterize the taxa and whose evolution we are trying to understand. These variables, like the data used to reconstruct phylogenies, need to be described and taken into account in a broad vision of the systematist's activity. The taxonomist thus has an ambitious set of specifications and a very wide range of know-how adapted to each study group.

3.3. A model group

The Eneopterinae subfamily was originally defined by Saussure as the "*légion des Eneopterites*" within the Orthoptera order and the Ensifera suborder, in the Grylloidea superfamily. These crickets have long been recognized as a distinct group in cricket taxonomy, either as a subfamily (see, for example, Kirby (1906), Chopard (1969), Desutter (1987) and Robillard and Desutter-Grandcolas (2008)) or tribe (see, for example, Chopard (1968) and Otte (1994)).

These crickets are studied for the analysis of their diversity and the origin of innovations in communication systems, particularly in island environments such as New Caledonia and in Southeast Asia.

Eneopterinae are crickets whose size varies between one and five centimeters. They have few exclusive characteristics that differentiate them from other crickets, but are recognizable by a set of morphological and anatomical features including:

– the general shape of the head: higher than it is wide when viewed from the front, with eyes that are laterally well developed;

– the spurs carried by the tibiae of the third pair of legs: four internal and four external spurs on the dorsal face, three internal and three external apical spurs, the median being the longest in both cases. Spines are present on the dorsal part of the posterior tibia, above and between the subapical spurs;

– the presence of an adhesive pad on the ventral side of the second tarsal segment: this feature, also present in other cricket clades, has, until recently, led to some confusion in the definition of the outlines of the subfamily;

– the tympanic organs on the tibia of the first pair of legs: in most Eneopterinae, the tympanic organs on the inner face are covered by a cuticle fold, which reveals the tympanic membrane by a thin longitudinal slit;

– the ovipositor of the female: well developed and laterally flattened.

Apart from these few common characteristics, Eneopterinae are very diverse in shape, size and coloration, as well as in the structure of their male genitalia, an organ widely used in cricket systematics and very different from one group of species to another (Robillard and Desutter-Grandcolas 2004b).

To focus on just a few contrasting examples of the phenotypes of these crickets (Figure 3.1), let us mention the *Lebinthus*, generally small, brown, stocky and brachypterous, reminiscent of wood crickets (*Nemobius sylvestris*); the genus *Nisitrus*, thin, bright yellow and black, with transparent wings; *Paranisitra*, a cousin of the latter, but apterous, mottled brown and yellow, with thin, elongated legs and a yellow band barring its face; or *Cardiodactylus*, often larger, with long elytra, usually decorated with specific patterns of yellow or white spots.

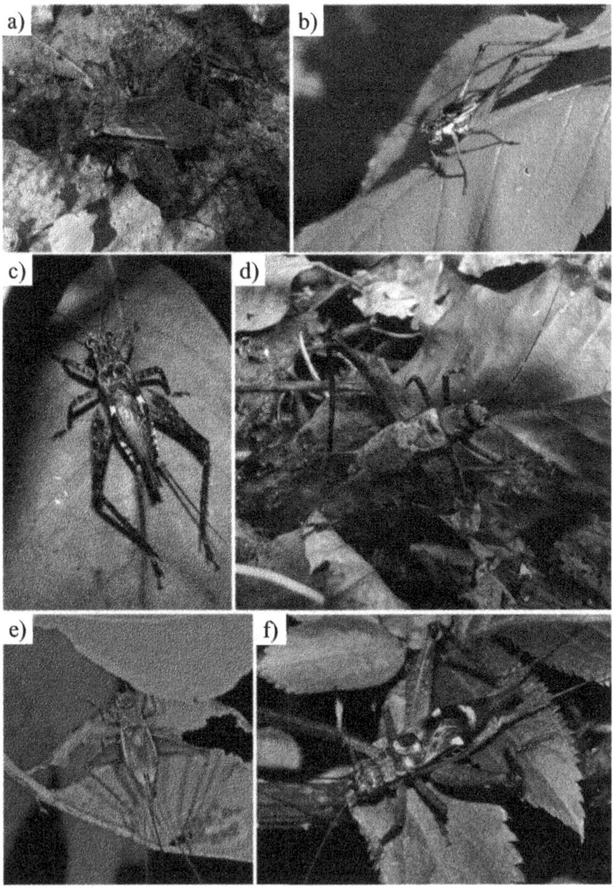

Figure 3.1. *Phenotypic diversity of Eneopterinae crickets. For a color version of this figure, see www.iste.co.uk/grandcolas/systematics.zip*

COMMENT ON FIGURE 3.1. – *a)* Lebinthus truncatipennis, *male (Indonesia, Java). b)* Nisitrus vittatus *(Singapore). c)* Cardiodactylus singapura, *male (Singapore, photo by Ming Kai Tan). d)* Centuriarus centurio, *male (Papua New Guinea). e)* Myara *sp., female (Australia, photo by David Rentz). f)* Cardiodactylus muria, *male (Indonesia, Java).*

Eneopterinae have a worldwide distribution, mostly in areas with a humid tropical climate, with a few exceptions such as the islands of the Indian Ocean (including Madagascar) and Hawaii. They live in more or less closed forest environments, but some species can be found on the edges of or in more open environments such as the African savanna, and some Australian species are even known to inhabit semi-desert environments. Some species are very well adapted to degraded or even semi-urban environments and frequently colonize crops, wastelands and gardens, and can be very common in parks in the middle of large conurbations such as Singapore or Kuala Lumpur (Robillard 2011; Robillard and Tan 2013; Tan and Robillard 2014).

In terms of habitat, most species live in leaf litter or on low plants or shrubs, and sometimes on the trunks of larger trees or in aerial litter. A few species occur in herbaceous environments, such as those of the genus *Xenogryllus*, particularly in African savannas (Jaiswara *et al.* 2019a). A common trait for many crickets, the species appears to have at least one active habitat, where males more or less expose themselves to emit their song, and a so-called "refuge" habitat, where individuals hide outside of active hours and pairs engage in long mating sequences. No formal habitat studies have yet been conducted to define the precise ecology of the species, but current data suggests that this aspect of their biology is important for their diversity. Similarly, if the period of cricket activity is defined as the time when the male cricket calls, then it can be said that some Eneopterinae are strictly diurnal (e.g. the genera *Eneoptera* and *Nisitrus*, which sing in the morning) or nocturnal (e.g. the genera *Agnotecous*, *Xenogrylllus*, *Ponca* and *Ligypterus*), while some species come out during twilight (certain *Cardiodactylus* and *Lebinthus*).

Within crickets, the Eneopterinae subfamily is a very relevant clade for studying questions related to diversification in island environments and the evolution of communication. There are three main reasons for this:

1) These crickets are a relatively manageable group in terms of the number of species; there are to date about 250 species described (see section

3.4.2), which is a lot for a group whose fine mechanisms of communication are being studied, but is still accessible for a diverse group of insects. This intermediate dimensioning allows the control of their taxonomy, and allows the development of studies on a rather large disciplinary panel. Similarly, for a clade of this size, it is relatively easy to reconstruct a phylogeny that is representative of the richness of the whole, and it is also easy to carry out studies at several taxonomic scales, both to deal with questions concerning closely related species, or even within populations, or by looking at more general questions of macroevolution or historical biogeography. The totality of the current fauna of these insects is still far from being known, although if studies are continued, then recent sampling efforts, combined with scientific collaborations, will make it possible to gain a thorough knowledge of the group.

2) Eneopterinae exhibit a high phenotypic and phylogenetic diversity, suggesting that their in-depth study can be seen as an adequate proxy in order to explore a large number of questions that can be generalized to crickets, or other clades of equivalent size and age. Likewise, their global distribution allows analyses at multiple nested scales, either to address general biogeographical questions or by focusing on a particular region.

3) Eneopterinae are among the most pronounced (or best known) cases of diversity associated with cricket communication, both in terms of the communication channels used and in terms of the structure of their sound-producing mechanisms, or the diversity of the signals emitted (see section 3.5).

3.4. Contribution of taxonomy for phylogenetic reconstructions and classification

3.4.1. *Monophyly*

The monophyly of Eneopterinae as a subfamily has been attested several times by morphological and molecular phylogenetic analyses at the scale of the subfamily (Robillard and Desutter-Grandcolas 2006; Nattier *et al.* 2011; Vicente *et al.* 2017), or more broadly at the scale of all crickets (Chintauan-Marquier *et al.* 2016). In light of the phylogenetic results, the Eneopterinae correspond to a clade within the Gryllidae family, as the sister group to the clade including the Gryllinae, the subfamily to which the archetypal species of crickets belong, that have served as models for many studies on acoustic communication and sexual selection (*Acheta domesticus*, *Gryllus bimaculatus*, *Teleogryllus oceanicus*, etc.).

3.4.2. *Recent taxonomic contributions*

In 2000, the Eneopterinae comprised 16 genera for about 120 species (Robillard and Desutter-Grandcolas 2004b). To date, the subfamily has significantly increased in size, as the Orthoptera Species File online database (Cigliano *et al.* 2020) lists 23 genera for 241 described species, which is six genera and 121 species more than 20 years ago (double). In fact, according to the material under study, there are at least 29 proven genera and about 300 species in the subfamily, including a significant proportion of species that are yet to be described and are already accessible.

These new taxa result from studies carried out by our team and by Russian and American colleagues, Otte (2007) having notably published descriptions of 39 new species of the genus *Cardiodactylus* (although a few of them have since been made synonymous (Robillard 2014; Robillard *et al.* 2014)).

While these figures speak for themselves in terms of the advances made in the group's taxonomic knowledge, they conceal many important steps covered by the generic term "taxonomic revision". Revising an existing genus can sometimes represent much more effort and time than describing a new one. Similarly, it can be very time-consuming to redescribe a species having a complex taxonomic history or to find the type specimens that will allow one to test a synonymy between two names that was previously proposed for a taxon. However, these thankless steps are necessary to clarify many ambiguous situations for sometimes key species of a taxonomic group. The studies on the taxonomy of Eneopterinae have thus led to the redescription or synonymization of certain species and to the definition of neotypes in the event of the proven loss of types for important species, whether they are type species of genera under study or very common species such as

Cardiodactylus novaeguineae or *Nisitrus vittatus* (Robillard and Ichikawa 2009; Robillard and Tan 2013; Robillard *et al.* 2013; Robillard 2014).

3.4.3. *Phylogeny and taxonomy*

The phylogeny of the group is quite stable through the different versions that have been published so far. Five monophyletic tribes can be distinguished, most of which have barely changed since the morphological phylogeny of 2004 (Robillard and Desutter-Grandcolas 2004b), and the modifications are mostly due to the contribution of molecular data and analytical methods than to taxonomic novelties (Figure 3.2): (1) the Eurepini tribe, endemic to Australia, is the sister group to the rest of the subfamily; (2) which then diverges into the Eneopterini, endemic to South and Central America; (3) the Nisitrini include species from Southeast Asia; (4) Xenogryllini are distributed in continental Asia and sub-Saharan Africa; and (5) Lebinthini, which are the most diverse, are present in both Asia and the archipelagos of the Western Pacific, as well as in South America.

The biogeographical history of the subfamily has recently been studied on the basis of Bayesian analyses and molecular dating. The topology of the tree, as presented in Figure 3.2, remains identical if we remove the branches corresponding to the taxa described after 1980, although only 16 species and 10 genera would then remain among the 62 species and 22 genera sampled by Vicente *et al.* (2017). Removing taxa *a posteriori* from a phylogenetic study is not in itself a test of topological stability. However, analyses including a lower sampling rate of the whole subfamily find identical or very close topology, as shown, for example, by the study by Anso *et al.* (2016), centered on New Caledonian taxa, or in the general phylogeny of crickets where each tribe of Eneopterinae is represented by only a few taxa (Chintauan-Marquier *et al.* 2016). The current phylogenetic structure is thus not affected by the important contribution of new taxa. However, the actual diversity of each of the clades is considered to be severely unbalanced to the disadvantage of the most diverse clades, such as the Lebinthini, whose number of species and genera has increased the most as a result of recent taxonomic revisions (Jaiswara *et al.* 2019b).

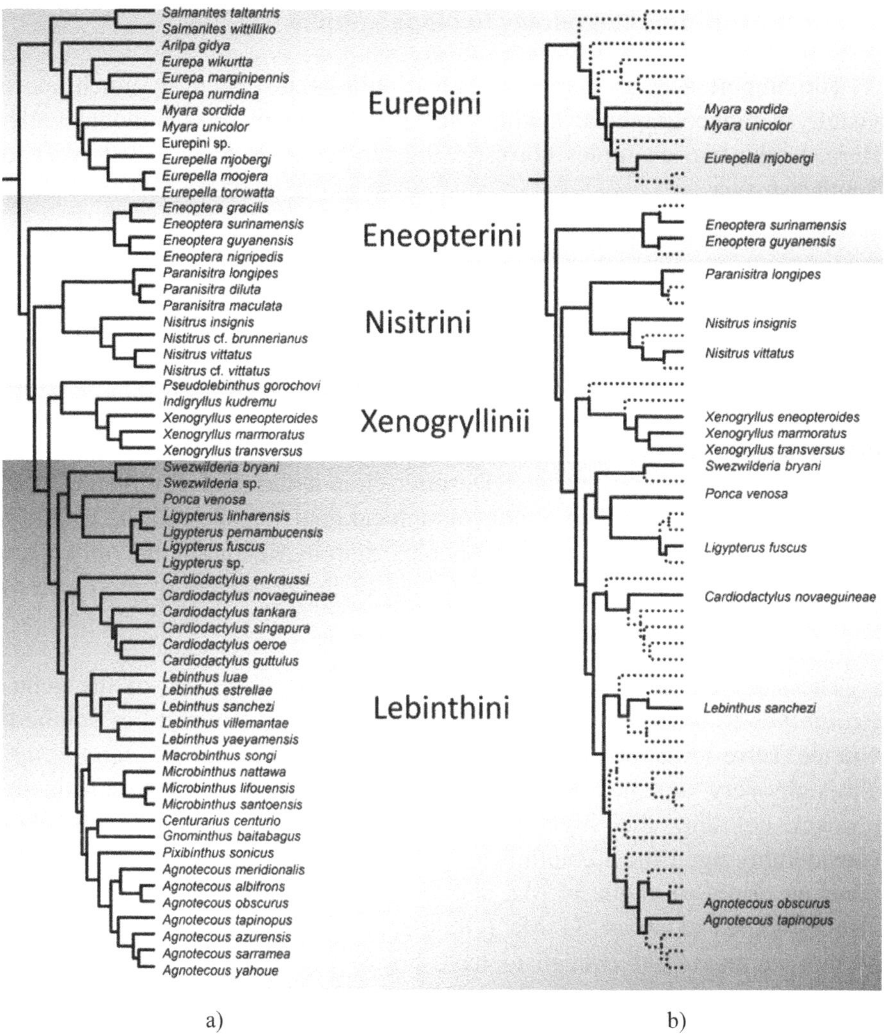

Figure 3.2. *Phylogenetic relationships of species of the Eneopterinae subfamily, reconstructed in Bayesian inference from eight molecular markers (COI, CytB, 12S, 16S, 12S, 18S, 28S and H3) (modified from Vicente et al. (2017)). For a color version of this figure, see www.iste.co.uk/grandcolas/systematics.zip*

COMMENT ON FIGURE 3.2. – *a) Complete tree. b) Tree only representing the names of species described before 1980 (branches leading to more recently described taxa are shown with dashed lines).*

3.5. Contribution of taxonomy to biogeography

The importance of taxonomic contributions is generally much more visible in biogeographical studies that are conducted on a regional scale. Here, I take two examples corresponding to work on New Caledonia and Southeast Asia.

3.5.1. *New Caledonia*

Eneopterinae are represented in New Caledonia through an endemic lineage of Lebinthini corresponding to the genus *Agnotecous* Saussure (Figure 3.3). In 2000, the genus had eight species, an already significant proportion of the Lebinthini diversity. Taxonomic revision led to the description of 12 new species (Desutter-Grandcolas and Robillard 2006; Robillard *et al.* 2010); then a biogeographical analysis allowed the genus to be dated at around 20 Ma with a probable origin in Melanesia, the only place in the Pacific region where representatives of these crickets were known to be present (Nattier *et al.* 2011).

The story becomes more complicated with the discovery of the genus *Pixibinthus* Robillard & Anso in 2016, in the "maquis minier" of southern Grande Terre (Anso *et al.* 2016). This new sister lineage of *Agnotecous*, although very distinct morphologically, ecologically and acoustically (cricket emitting the highest frequencies known to date, 28 kHz), considerably aged the Lebinthini origin on Grande Terre, since the dating gives an origin of about 34 Ma, shortly after the re-emersion of the island also estimated at about 34 Ma (Grandcolas *et al.* 2008). No territory in Melanesia seems old enough to have hosted the ancestor of this lineage. In addition, many taxa in New Guinea that may be related to the New Caledonian lineage have been discovered in the meantime (Robillard *et al.* 2016), and another new genus very similar to *Pixibinthus* has been discovered even more recently in Queensland: *Julverninthus* Robillard & Su is the first genus of Lebinthini that is endemic to Australia (Robillard and Su 2018).

The question of the origin in time and space of these crickets in New Caledonia goes far beyond the framework of Grande Terre. It is a regional question, whose resolution will depend on future phylogenetic results and analyses of historical biogeography. However, the problem is undeniably linked to the many very recent advances in taxonomy and the discovery of new key taxa.

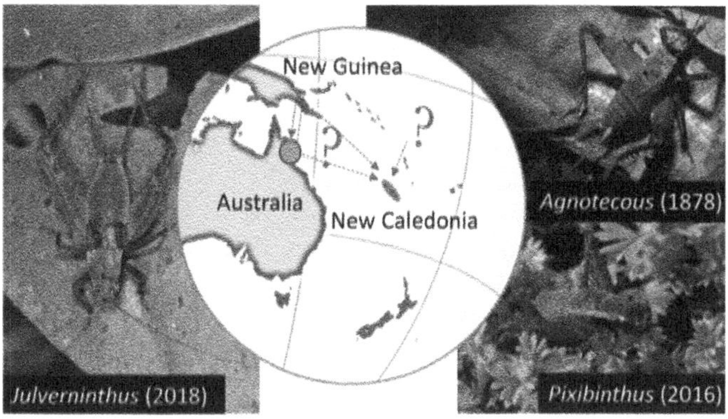

Figure 3.3. *Schematic representation of the geographical situation of New Caledonia and the Lebinthini lineages present in the region. For a color version of this figure, see www.iste.co.uk/grandcolas/systematics.zip*

3.5.2. *Southeast Asia*

The second example showing the importance of taxonomy in phylogenetic and biogeographical analyses on a regional scale is the genus *Cardiodactylus*, which has recently been the subject of numerous taxonomic studies. Prior to 2007, only nine species of these crickets with a confusing taxonomy were known (Robillard and Desutter-Grandcolas 2004b). A series of recent studies has led to the revision of the genus and the description of 73 new species (see, for example, Otte (2007), Robillard (2009, 2011, 2014), Robillard *et al.* (2014) and Tan and Robillard (2014)). In light of this diversity, a historical biogeographic study has demonstrated that the settlement of these crickets in Southeast Asia is the result of multiple colonization events, originating in the islands of the Western Pacific, with New Guinea as the point of entry into Asia (Dong *et al.* 2018). This study is entirely based on recent taxonomic work and thus could not have been conducted a few years earlier without it. Indeed, before 2007, only two species of the genus *Cardiodactylus* were known from this region (Figure 3.4): *C. guttulus* Matsumura, known from the Ryūkyū islands and Taiwan, and which was absent from taxonomic catalogs before being rediscovered and redescribed in 2009 (Robillard and Ichikawa 2009), and *C. novaeguineae* (Haan), which is a widely distributed species found on the coasts of most islands in the western Pacific and as far west as New Guinea.

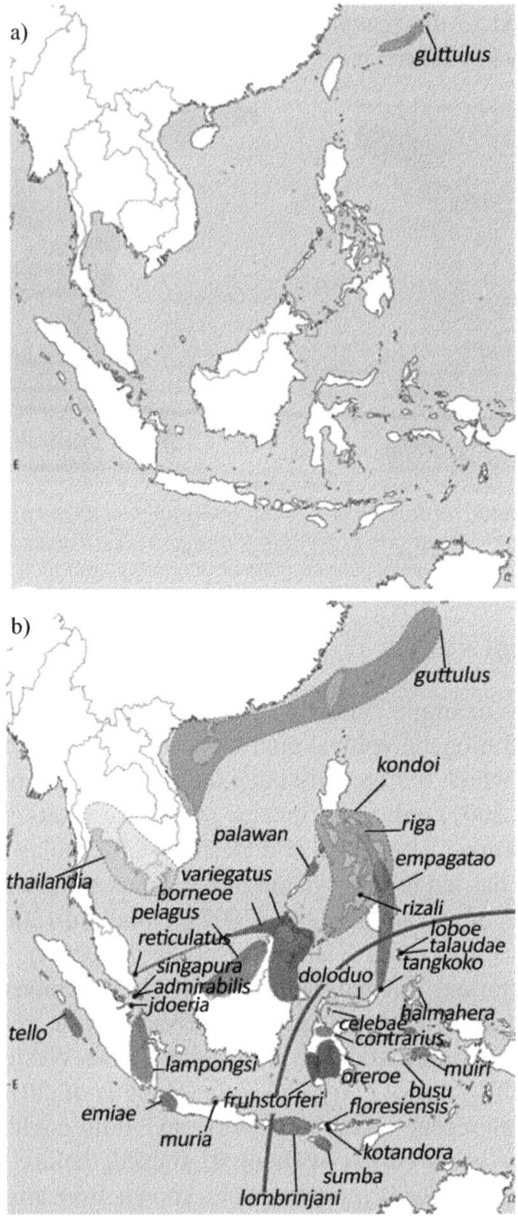

Figure 3.4(A). *Schematic representation of the distribution areas of* Cardiodactylus *crickets in Southeast Asia with respect to their phylogenetic relationships. For a color version of this figure, see www.iste.co.uk/grandcolas/systematics.zip*

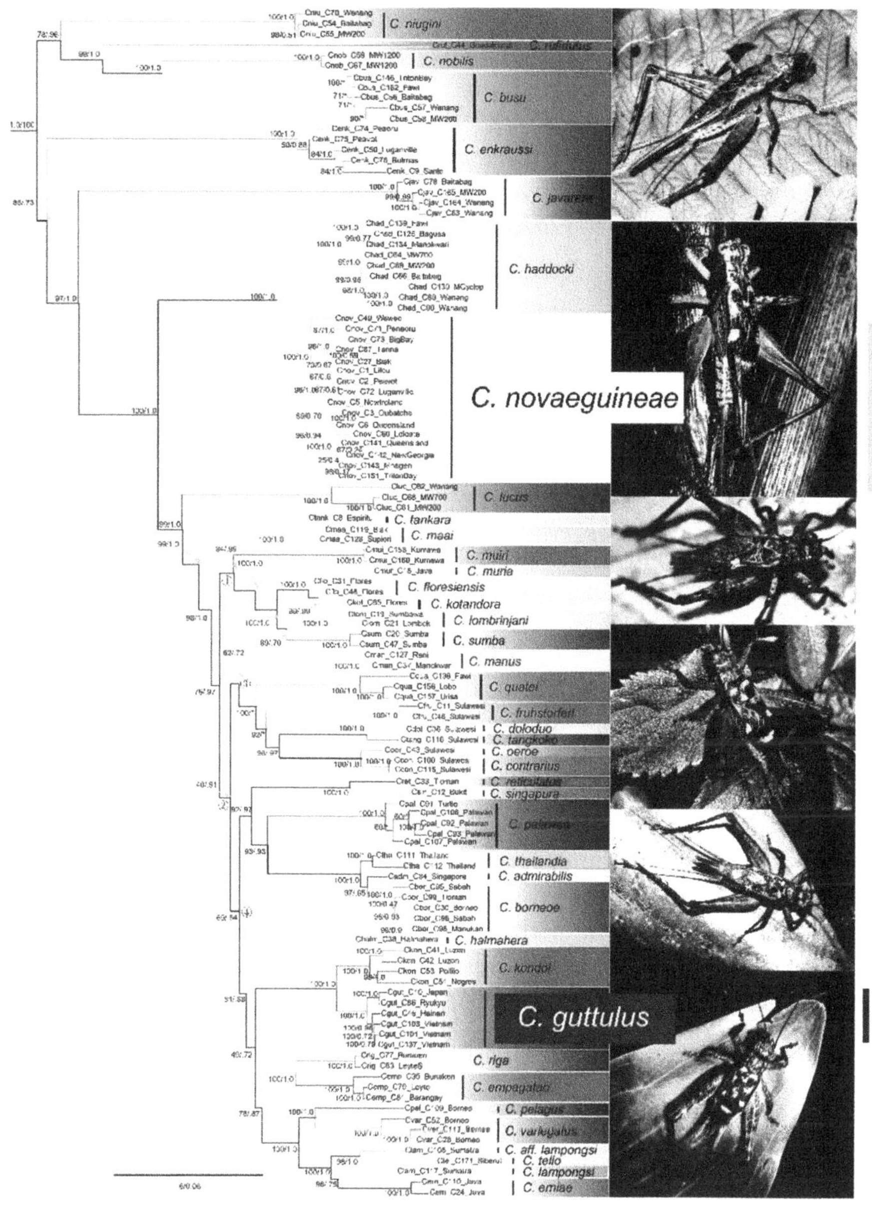

c)

Figure 3.4(B). *Schematic representation of the distribution areas of* Cardiodactylus *crickets in Southeast Asia with respect to their phylogenetic relationships (continued). For a color version of this figure, see www.iste.co.uk/grandcolas/systematics.zip*

COMMENT ON FIGURE 3.4. – *a) Distribution map prior to 2007 (only one species recorded in the region,* C. novaeguineae *having only recently been reported west of New Guinea). b) Distribution map of current species (modified from Dong* et al. *(2018)). c) Maximum likelihood inferred molecular phylogeny indicating the position of the only two known species in the region prior to 2007, shown to the right with vertical colored lines (modified from Dong* et al. *(2018)).*

3.6. Taxonomic exploration and evolution of species traits

Although taxonomic exploration makes phylogenetic analyses, classification systems and biogeographical interpretations more reliable, it also has a role to play in the knowledge of the traits of the species themselves. Beyond the act of naming and describing taxa, taxonomy provides access to a wealth of information for the entire scientific community. It samples the diversity of species well beyond model or archetypal species, which offers many opportunities for unexpected discoveries.

In the case of Eneopterinae, the discovery of new taxa is motivated by the study of the evolution of communication. In the early 2000s, the morphological diversity of these crickets suggested an underlying acoustic diversity, to which was added the knowledge of a few species showing very particular cases of emitted signals and sound-production structures. For example, the song of the species *Eneoptera guyanensis* Chopard has two very distinct parts, produced alternately with a double stridulatory file, which is a unique case in crickets (Desutter-Grandcolas 1998). A thorough study of the behavior of this species has revealed that this song is generated by complex movements of the elytra (Robillard and Desutter-Grandcolas 2011) and that it also has ultrasonic components throughout its duration (Robillard *et al.* 2015).

In 2004, when the Lebinthini tribe was established (Robillard and Desutter-Grandcolas 2004b), only four species out of the then-known 20 had an acoustic recording. Analysis of these early data revealed that these species used high-frequency signals, and that unlike other crickets, their dominant frequency corresponded to a harmonic of the fundamental frequency (Robillard and Desutter-Grandcolas 2004a; Robillard *et al.* 2007).

As indicated above, the Lebinthini are the tribe of Eneopterinae that has increased the most in the number of taxa over the last two decades, from about 15 species in three genera to nearly 150 species in 19 genera (and even more if we count the taxa that has been discovered, but are yet to be described). Very quickly, it turned out that all the new species that were discovered belonging to this clade produced high-frequency signals. What makes things interesting is that these new species are not just units added to a number of species, nor are they simple replicas of already known species. Each new species discovered is the result of a unique evolutionary history that brings its share of phenotypic novelties, including acoustic signals. All of these characteristics are added to what is known from the panel of possibilities realized for the study group, which is invaluable for the study of evolution. Each new taxon brings a unique mosaic of evolutionary possibilities, expressed through the selection filter in a given temporal and geographical context. Thus, over the course of the descriptions, it has been possible to document signals in the Lebinthini whose dominant frequency may be the second, third, fourth or even fifth harmonic peak, with frequency values far beyond the limit of ultrasound. The record for the highest known frequency of a cricket song is currently held by the New Caledonian species *Pixibinthus sonicus* Anso & Robillard, which emits a long trill consisting of thousands of syllables at a frequency of 28 kHz.

Taxonomic exploration does not only describe the diversity of phenotypes, it also contributes by raising new questions. We can suddenly discover a species which, although similar in many aspects to other species of the group to which it belongs, presents a singular characteristic that calls into question a whole panoply of our knowledge. These particular characteristics can be as diverse as a new structure or behavior, a gene with novel properties (if we have access to this level of information), a pivotal phylogenetic position, a geographical location that opens up new biogeographical interpretations, or the age of a new fossil that calls into question the entire history of the group. Evolutionary biology, like many other fields, is also a dynamic discipline because it is based on such discoveries that constantly feed it.

It is sometimes the sum of this new knowledge that will raise questions. This was the case for the Eneopterinae crickets. A species with an astonishing song at high frequencies is just a curious thing. When several dozen species share this innovation, or each exhibit it in its own way, we begin to wonder how this impacts the communication system. The fact that the frequency of the cricket song differs from one species to another is nothing exceptional. However, knowing whether the frequency is doubled, tripled or quintupled because of the use of dominant harmonic frequencies is intriguing.

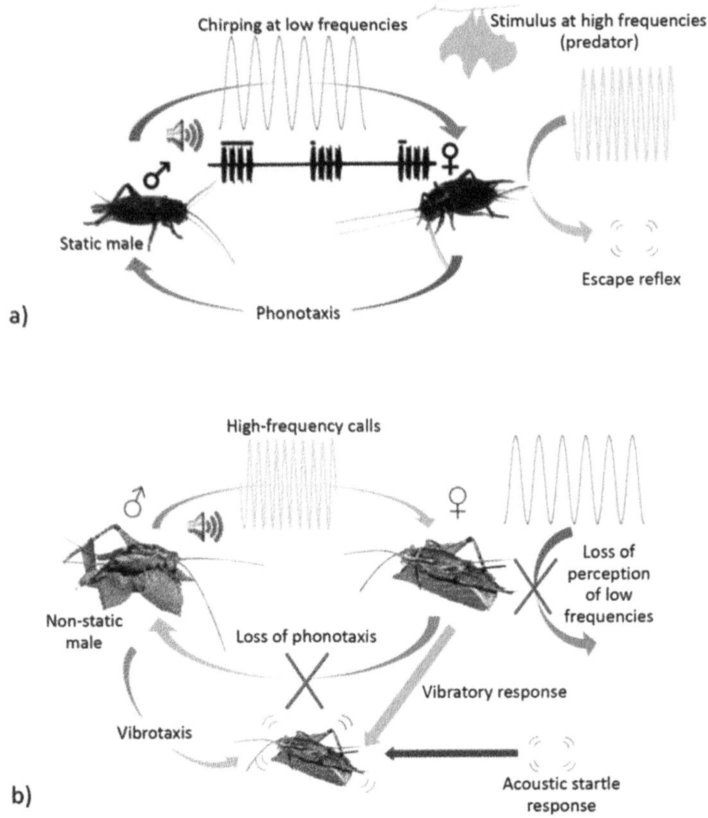

Figure 3.5. *Cricket intraspecific communication systems. For a color version of this figure, see www.iste.co.uk/grandcolas/systematics.zip*

COMMENT ON FIGURE 3.5. – *a) Communication system in archetypal species (Gryllinae), including low-frequency calling and phonotactic response of the female, and in parallel, presence of a startle response when exposed to high-frequency stimuli from bat predators. b) Lebinthini communication system including high-frequency calling of the male and vibratory response of the female, acquired by sensory exploitation of the startle response present in other crickets.*

Indeed, in the Gryllinae (archetypal crickets), which use songs at low frequencies (about 5 kHz), the perception of singing is operated by an auditory neuron called AN1, sensitive to frequencies below 10 kHz (Gerhardt and Huber 2002). In these species, stimuli at high frequencies (> 10 kHz) are perceived by another auditory neuron (AN2) and provoke a violent flight reflex related to predation by bats, some of which use their echolocation (ultrasound) signals to locate their prey (see, for example, Weber and Thorson (1989), Wyttenbach *et al.* (1996)). In other words, it is quite surprising that the Lebinthini use communication signals that are generally perceived by female crickets, as predator signals. This realization was the basis for the study to understand how these crickets communicate using these particular signals.

These studies, which are currently ongoing, have led to the discovery of a new communication system in these crickets (Figure 3.5) which was probably acquired by sensory exploitation of the acoustic startle response existing in other crickets, a rarely encountered evolutionary situation (ter Hofstede *et al.* 2015). Remarkably, the signal inducing escape – high-frequency sound – would have been incorporated here into the intraspecific communication system in order to form a new and totally original communication system: the female, which has even lost the perception of low frequencies, does not orient herself towards the male when she hears his high-frequency calling song (absence of phonotaxis); instead, she emits a vibratory response that guides the male to her through the plant substrate. This system of communication is supposed to have played a major role in the diversification of the Lebinthini. However, it is based on the acquisition of a single trait shared by all these crickets: the elytra of the males have acquired the ability to vibrate at higher frequencies by amplifying one of the harmonics of the frequency spectrum generated by singing, a phenomenon called "harmonic hopping".

Taxonomic exploration of these crickets has made this discovery possible, but more importantly, it allows us to characterize each stage of the system and its inherent diversity. Initially described in three species, the new Lebinthini system has now been observed in many others thanks to the recording of calling songs of dozens of species with different morphologies and habitats. Taxonomy also allows the discovery of new lineages within these crickets that call into question the universality of this communication system, such as the discovery of certain species that only sing from the forest litter where the females never respond to the male with vibrations (Robillard, personal observations). Similarly, the taxonomic revision of the South American Eneopterinae has led to a better characterization of the neotropical lineage of Lebinthini (Robillard and Desutter-Grandcolas 2005). Recently rediscovered in Panama, the species *Ponca hebardi* Robillard was found to have the same communication system as species of the Asian lineage, but its study also revealed that males also interact with each other by both responding with high-frequency sounds and producing vibrations mimicking the responses of females, which corresponds to the first case of multimodal interactions ever recorded in male crickets (Benavides-Lopez *et al.* 2020). This observation raises many questions, including whether the case of *Ponca hebardi* is a complexification of the system existing in other Lebinthini, or whether these male–male interactions exist in other species. In any case, it puts forward many questions as to the reproductive strategies used by this species and the origin of such behaviors, revealed thanks to the knowledge provided by the taxonomic study.

Finally, one of the fascinating aspects of the evolution of acoustic communication in crickets is its outright disappearance from many of today's lineages. Such losses of acoustic communication resulting from multiple evolutionary convergences are particularly known in the Eneopterinae crickets, to the point that each tribe within the subfamily has a different example. In some lineages, the ability to produce sounds disappeared, but the wings and the ability to fly persisted, as in the genus *Swezwilderia*; in others, the wings were entirely lost, as in *Paranisitra*, while in the species *Pseudolebinthus lunipterus* Salazar *et al.*, both the stridulatory apparatus and auditory tympana have been lost (Salazar *et al.* 2020) within an enigmatic genus of African species that produce high-frequency signals using highly modified elytra. All these discoveries, and the cases of study of complex phenomena that result from them, obviously stem from taxonomic work.

3.7. Conclusion

The example of the Eneopterinae crickets has demonstrated that taxonomy, descriptive approaches and natural history as a whole are paramount in the study of evolution, both to reconstruct phylogenetic relationships at all levels, to trace the biogeographical history of lineages and to explore the diversity of taxa traits.

To conclude this chapter, I would like to encourage everyone to consider the usefulness of these approaches in their own evolutionary research. What can taxonomy and natural history do for you today?

The answer will depend on the context and the taxonomic group in which each person works, and therefore on past efforts to consolidate knowledge about the group in question.

For the luckiest, the answer will be "nothing, because everything is clear, everything has been described and the data is accessible, available, and we can move on to other studies building on this knowledge". Congratulations, you are working on a perfectly documented group, both in the present, in space (has that small mountain in the depths of Asia been documented too?) and in the fossil record (there is now room for doubt).

A few cases certainly exist for particularly well-studied vertebrates, such as species of the genus *Homo*, for example. However, are we not still discovering new species that are related to us, such as *Homo luzonensis*, recently discovered in northern Luzon in the Philippines (Detroit *et al.* 2019)? How about elephants? Giraffes? So, no, the taxonomic debates are far from over, even for such well-studied groups, as recent studies show (see, for example, Bercovitch *et al.* (2017)). In many cases, particularly for large vertebrates, there are certainly not many species that are alive today that are still being described, but taxonomy still has a role to play in the discovery of fossil species, as well as in the sometimes-difficult questions aimed at delineating the number of taxa that can be recognized among a given set of current populations.

However, in 99% of cases, taxonomy still has many services to render to the scientific community and society. Here are a few subsidiary questions to help you realize this: do you know precisely how many species make up the group you are studying, at the genus, family or phylum level? What is the distribution of these species and the groups they constitute? If you look at a map of these distributions, is it complete or are there obvious gaps?

Now, take a well-known species, such as the type species of the genus whose phylogenetic relationships you have reconstructed, or the one whose genome you recently sequenced. Does this species have a clear and unambiguous description? Does it have a single name or five, or ten more or less well-defined synonyms? And even more simply, when was the last taxonomic revision of the group to which it belongs? Do we know its habitat, its sexual behavior, its means of communication? And what about the species that is closest to it in phylogeny?

In some cases, we may feel that these issues were all resolved decades ago and that it is time to move on. However, the concepts, methods and knowledge of taxonomy are far from what they might have been 50 years ago. Systematic researchers did not have the same techniques and approaches that are used today, so if research is resumed, some surprises might be revealed in light of new data and techniques that are constantly advancing, in order to associate ever richer and more precise information with taxon concepts. Very often, a system that is stable at a time *t* will no longer be so when observing another level of reading, whether it be molecular, genomic or phylogenetic. Similarly, morphological traits characterized as "variable" in the past may, for example, reveal the presence of cryptic species if sophisticated geometric morphometry methods are used, even in the absence of genetic material. Thus, even in the presence of a finite sample, taxonomy has a role to play.

Entire sections of biodiversity can thus be hidden behind groups that are believed to be "well known", and it is not only a question of describing new species with the sole (and probably futile) objective of listing everything, but also of making sure that what is believed to be known is what appears to be known.

3.8. Acknowledgments

I would like to thank the many collaborators, interns, doctoral students, and occasional or long-time colleagues who made it possible to develop the studies mentioned in this chapter, and in particular Frédéric Legendre for his critical reading of a preliminary version of the manuscript. I would also like to thank ISYEB and the MNHN for their indispensable financial support for the taxonomy work.

3.9. References

Alfred, J. and Baldwin, I.T. (2015). New opportunities at the wild frontier. *eLife*, 4, 1–4.

Anso, J., Barrabe, L., Desutter-Grandcolas, L., Jourdan, H., Grandcolas, P., Dong, J., and Robillard, T. (2016). Old lineage on an old island: *Pixibinthus*, a new cricket genus endemic to New Caledonia shed light on gryllid diversification in a hotspot of biodiversity. *PLOS One*, 11(3), e0150920 [Online]. Available: https://journals.plos.org/plosone/article?id=10.1371/journal.pone.0150920.

Baum, D.A. and Smith, S.D. (2012). *Tree Thinking: An Introduction to Phylogenetic Biology*. Roberts and Company Publishers, Greenwood Village.

Benavides-Lopez, J.L., ter Hofstede, H., and Robillard, T. (2020). Novel system of communication in crickets originated at the same time as bat echolocation and includes male–male multimodal communication. *The Science of Nature*, 107(1), 9 [Online]. Available: http://link.springer.com/10.1007/s00114-020-1666-1.

Bercovitch, F.B., Berry, P.S.M., Dagg, A., Deacon, F., Doherty, J.B., Lee, D.E., Mineur, F., Muller, Z., Ogden, R., Seymour, R., Shorrocks, B., and Tutchings, A. (2017). How many species of giraffe are there?. *Current Biology*, 27(4), R136–R137 [Online]. Available: http://dx.doi.org/10.1016/j.cub.2016.12.039.

Brenowitz, E.A. and Zakon, H.H. (2015). Emerging from the bottleneck: Benefits of the comparative approach to modern neuroscience. *Trends in Neurosciences*, 38(5), 273–278 [Online]. Available: http://www.ncbi.nlm.nih.gov/pubmed/25800324 [Accessed 19 January 2020].

Chintauan-Marquier, I.C., Legendre, F., Hugel, S., Robillard, T., Grandcolas, P., Nel, A., Zuccon, D., and Desutter-Grandcolas, L. (2016). Laying the foundations of evolutionary and systematic studies in crickets (Insecta, Orthoptera): A multilocus phylogenetic analysis. *Cladistics*, 32(1), 54–81.

Chopard, L. (1968). Pars 12. Fam. Gryllidae: Subfam. Mogoplistinae, Myrmecophilinae, Scleropterinae, Cachoplistinae, Pteroplistinae, Pentacentrinae, Phalangopsinae, Trigonidiinae, Eneopterinae. Fam. Oecanthidae, Gryllotalpidae. In *Orthopterorum Catalogue*, Beier, M. (ed.). Dr. W. Junk Publishers, The Hague.

Chopard, L. (1969). *The Fauna of India and Adjacent Countries: Orthoptera, Volume 2*. Grylloidea Baptist Mission Press, Calcutta.

Cigliano, M.M., Braun, H., Eades, D.C., and Otte, D. (2020). Orthoptera Species File. Version 5.0/5.0 [Online]. Available: http://orthoptera.speciesfile.org/HomePage/Orthoptera/HomePage.aspx.

Desutter, L. (1987). Structure et évolution du complexe phallique des Gryllidea (Orthoptera) et classification des genres néotropicaux de Grylloidea (1ère partie). *Annales de la Société Entomologique de France (Nouvelle Série)*, 23, 213–239.

Desutter-Grandcolas, L. (1998). Broad-frequency modulation in cricket (Orthoptera, Grylloidea) calling songs: Two convergent cases and a functional hypothesis. *The Canadian Journal of Zoology*, 76, 2148–2163.

Desutter-Grandcolas, L. and Robillard, T. (2006). Phylogenetic systematics and evolution of Agnotecous in New Caledonia (Orthoptera: Grylloidea, Eneopteridae). *Systematic Entomology*, 31(1), 65–92.

Détroit, F., Mijares, A.S., Corny, J., Daver, G., Zanolli, C., Dizon, E., Robles, E., Grün, R., and Piper, P.J. (2019). A new species of *Homo* from the Late Pleistocene of the Philippines. *Nature*, 568(7751), 181–186 [Online]. Available: https://doi.org/10.1038/s41586-019-1067-9.

Dong, J., Kergoat, G.J., Vicente, N., Rahmadi, C., Xu, S., and Robillard, T. (2018). Biogeographic patterns and diversification dynamics of the genus *Cardiodactylus* Saussure (Orthoptera, Grylloidea, Eneopterinae) in Southeast Asia. *Molecular Phylogenetics and Evolution*, 129 (June), 1–14 [Online]. Available: https://doi.org/10.1016/j.ympev.2018.06.001.

Gerhardt, H.C. and Huber, F. (2002). *Acoustic Communication in Insects and Anurans*. The University of Chicago Press, Chicago and London.

Grandcolas, P., Deleporte, P., and Desutter-Grandcolas, L. (1994). Why to use phylogeny in evolutionary ecology? *Acta Oecologica*, 15, 661–673.

Grandcolas, P., Deleporte, P., and Desutter-Grandcolas, L. (1997). Testing evolutionary processes with phylogenetic patterns: Test power and test limitations. In *The Origin of Biodiversity in Insects: Phylogenetic Tests of Evolutionary Scenarios*, Grandcolas, P. (ed.). Mémoires du Muséum national d'histoire naturelle, Paris.

Grandcolas, P., Murienne, J., Robillard, T., Desutter-Grandcolas, L., Jourdan, H., Guilbert, E., and Deharveng, L. (2008). New Caledonia: A very old Darwinian island?. *Philosophical Transactions of the Royal Society B*, 363(1508) [Online]. Available: https://royalsocietypublishing.org/doi/full/10.1098/rstb.2008.0122.

ter Hofstede, H.M., Schöneich, S., Robillard, T., Hedwig, B. (2015). Evolution of a communication system by sensory exploitation of startle behavior. *Current Biology*, 25(24), 3245–3252.

Jaiswara, R., Dong, J., Ma, L., Yin, H., and Robillard, T. (2019a). Taxonomic revision of the genus *Xenogryllus* Bolívar, 1890 (Orthoptera, Gryllidae, Eneopterinae, Xenogryllini). *Zootaxa*, 4545(3), 301–338.

Jaiswara, R., Dong, J., and Robillard, T. (2019b). Phylogenetic relationships in the cricket tribe Xenogryllini (Orthoptera, Gryllidae, Eneopterinae) and description of the Indian genus *Indigryllus* gen. nov. *Journal of Zoological Systematics and Evolutionary Research*, 57(4) [Online]. Available: https://onlinelibrary.wiley.com/doi/abs/10.1111/jzs.12298.

Kirby, W.F. (1906). *A Synonymic Catalogue of Orthoptera*. British Museum, London.

Lemey, P., Salemi, M., and Vandamme, A.-M. (2009). *The Phylogenetic Handbook: A Practical Approach to Phylogenetic Analysis and Hypothesis Testing*. Cambridge University Press, Cambridge.

Losos, J.B. (2009). *Lizards in an Evolutionary Tree: The Ecology of Adaptive Radiation in Anoles*. University of California Press, Berkeley.

Losos, J.B. (2011). Seeing the forest for the trees: The limitations of phylogenies in comparative biology. *The American Naturalist*, 177(6), 709–727.

Manger, P.R., Cort, J., Ebrahim, N., Goodman, A., Henning, J., Karolia, M., Rodrigues, S.L., and Štrkalj, G. (2008). Is 21st century neuroscience too focussed on the rat/mouse model of brain function and dysfunction? *Frontiers in Neuro-Anatomy*, 2(November), 1–7.

Miralles, A., Raymond, M., and Lecointre, G. (2019). Empathy and compassion toward other species decrease with evolutionary divergence time. *Scientific Reports*, 9(1), 1–8.

Nattier, R., Robillard, T., Desutter-Grandcolas, L., Couloux, A., and Grandcolas, P. (2011). Older than New Caledonia emergence? A molecular phylogenetic study of the eneopterine crickets (Orthoptera: Grylloidea). *Journal of Biogeography*, 38(11), 2195–2209.

Otte, D. (1994). *Orthoptera Species File 1. Crickets (Grylloidea)*. Orthopterists's Society and the ANSP, Philadelphia.

Otte, D. (2007). New species of *Cardiodactylus* from the western Pacific region (Gryllidae: Eneopterinae). *Proceedings of the Academy of Natural Sciences of Philadelphia*, 156, 341–400.

Robillard, T. (2009). Eneopterinae crickets (Insecta, Orthoptera, Grylloidea) from Vanuatu. *Zoosystema*, 31(3), 577–618 [Online]. Available: http://www.bioone.org/doi/abs/10.5252/z2009n3a11.

Robillard, T. (2011). New *Cardiodactylus* species from unsuspected places in Southeast Asia (Orthoptera, Grylloidea, Eneopterinae). *Zootaxa*, 26(2909), 14–26.

Robillard, T. (2014). Review and revision of the century-old types of *Cardiodactylus* crickets (Grylloidea, Eneopterinae, Lebinthini). *Zoosystema*, 36(1), 101–125.

Robillard, T. and Desutter-Grandcolas, L. (2004a). High-frequency calling in Eneopterinae crickets (Orthoptera, Grylloidea, Eneopteridae): Adaptive radiation revealed by phylogenetic analysis. *Biological Journal of the Linnean Society*, 83(4), 577–584.

Robillard, T. and Desutter-Grandcolas, L. (2004b). Phylogeny and the modalities of acoustic diversification in extant Eneopterinae (Insecta, Orthoptera, Grylloidea, Eneopteridae). *Cladistics*, 20(3), 271–293.

Robillard, T. and Desutter-Grandcolas, L. (2005). A revision of Neotropical Eneopterinae crickets (Orthoptera, Grylloidea, Eneopteridae) with a phylogenetic discussion. *Insect Systematics & Evolution*, 25(4), 411–435 [Online]. Available at: https://doi.org/10.1163/187631204788912427.

Robillard, T. and Desutter-Grandcolas, L. (2006). Phylogeny of the cricket subfamily Eneopterinae (Orthoptera, Grylloidea, Eneopteridae) based on four molecular loci and morphology. *Molecular Phylogenetics and Evolution*, 40(3), 643–661. DOI: 0.1016/j.ympev.2005.10.019.

Robillard, T. and Desutter-Grandcolas, L. (2008). Clarification of the taxonomy of extant crickets of the subfamily Eneopterinae (Orthoptera: Grylloidea; Gryllidae). *Zootaxa*, 68(1789), 66–68.

Robillard, T. and Desutter-Grandcolas, L. (2011). The complex stridulatory behavior of the cricket *Eneoptera guyanensis* Chopard (Orthoptera: Grylloidea: Eneopterinae). *Journal of Insect Physiology*, 57(6), 694–703 [Online]. Available: http://dx.doi.org/10.1016/j.jinsphys.2011.02.005.

Robillard, T. and Ichikawa, A. (2009). Redescription of two *Cardiodactylus* species (Orthoptera, Grylloidea, Eneopterinae): The supposedly well-known *C. novaeguineae* (Haan, 1842), and the semi-forgotten *C. guttulus* (Matsumura, 1913) from Japan. *Zoological Science*, 26(12), 878–891.

Robillard, T. and Su, Y.N. (2018). New lineages of Lebinthini from Australia (Orthoptera: Gryllidae: Eneopterinae). *Zootaxa*, 4392(2), 241.

Robillard, T. and Tan, M.K. (2013). A taxonomic review of common but little known crickets from Singapore and the Philippines (Insecta: Orthoptera: Eneopterinae). *Raffles Bulletin of Zoology*, 61(2) [Online]. Available: https://www.researchgate.net/publication/260414740_A_taxonomic_review_of_common_but_little_known_crickets_from_Singapore_and_the_Philippines_Insecta_Orthoptera_Eneopterinae.

Robillard, T., Grandcolas, P., and Desutter-Grandcolas, L. (2007). A shift toward harmonics for high-frequency calling shown with phylogenetic study of frequency spectra in Eneopterinae crickets (Orthoptera, Grylloidea, Eneopteridae). *Canadian Journal of Zoology*, 85(12), 1264–1275.

Robillard, T., Nattier, R., and Desutter-Grandcolas, L. (2010). New species of the New Caledonian endemic genus *Agnotecous* (Orthoptera, Grylloidea, Eneopterinae, lebinthini). *Zootaxa*, 35(2559), 17–35.

Robillard, T., Yap, S., and Yngente, M.V. (2013). Systematics of cryptic species of Lebinthus crickets in Mount Makiling (Grylloidea, Eneopterinae). *Zootaxa*, 3693(1), 49–63.

Robillard, T., Gorochov, A.V., Poulain, S., and Suhardjono, Y.R. (2014). Revision of the cricket genus *Cardiodactylus* (Orthoptera, Eneopterinae, Lebinthini): The species from both sides of the Wallace line, with description of 25 new species. *Zootaxa*, 3854(1), 1–104.

Robillard, T., ter Hofstede, H.M., Orivel, J., and Vicente, N.M. (2015). Bioacoustics of the Neotropical Eneopterinae (Orthoptera, Grylloidea, Gryllidae). *Bioacoustics, the International Journal of Animal Sound and Its Recording*, 24(2) [Online]. Available: https://www.tandfonline.com/doi/abs/10.1080/09524622.2014.996915.

Robillard, T., Dong, J., Legendre, F., and Agauvoa, S. (2016). The brachypterous Lebinthini crickets from Papua New Guinea, with description of two new genera and four new species (Orthoptera: Gryllidae: Eneopterinae). In *Insects of Mount Wilhelm, Papua New Guinea*, Robillard, T., Legendre, F., Villemant, C. and Leponce, M. (eds). Mémoires du Muséum national d'histoire naturelle, Paris.

Salazar, K., Murphy, R.J., Guillaume, M., Nattier, R., and Robillard, T. (2020). *Pseudolebinthus lunipterus* sp. nov.: A striking deaf and mute new cricket from Malawi (Orthoptera, Gryllidae, Eneopterinae). *PeerJ*, 8, e8204 [Online]. Available: https://peerj.com/articles/8204/.

Tan, M.K. and Robillard, T. (2014). A new species of *Cardiodactylus* (Orthoptera: Grylloidea: Eneopterinae) from Singapore. *Zootaxa*, 3764(3), 364–376.

Troudet, J., Grandcolas, P., Blin, A., Vignes-Lebbe, R., and Legendre, F. (2017). Taxonomic bias in biodiversity data and societal preferences. *Scientific Reports*, 7(1), 1–14.

Vicente, N., Kergoat, G.J., Dong, J., Yotoko, K., Legendre, F., Nattier, R., and Robillard, T. (2017). In and out of the Neotropics: Historical biogeography of Eneopterinae crickets. *Journal of Biogeography*, 44(10) [Online]. Available: https://onlinelibrary.wiley.com/doi/abs/10.1111/jbi.13026.

Weber, T. and Thorson, J. (1989). Phonotactic behavior of walking crickets. In *Cricket Behavior and Neurobiology*, Huber, F., Moore, T.E., and Loher, W. (eds). Cornell University Press, Ithaca.

Wyttenbach, R.A., May, M.L., and Hoy, R.R. (1996). Categorical perception of sound frequency by crickets. *Science*, 273, 1542–1544.

4

Systematics in the (Post)genomic Era: A Look at the Drosophila Model

> At the risk of seeming to confuse science and science fiction, I will end this review with a few undisguised speculations. At this stage in the history of biology, considerable effort is devoted to the study of the genetics of development and shape. At some point in the future, say in five, 10 or 20 years, a door will be opened on the genetic machinery of wing venation, dorsocentral setae, cerci and tagmosis – at this point, morphological and molecular data will have a common genetic denominator. Since so much development research has been done on Drosophila, entomologists will be closer to this door than anyone else. (Berlocher 1984)

4.1. Drosophila: a star of genetics but a systematic nebula

The process of speciation is often gradual, culminating in the evolution of complete reproductive isolation and the fixation of genetic differences between emerging species. Drawing lines along this continuum has long been an important challenge for the systematist. Recent advances in genome sequencing have provided an incomparably rich source of traits with objective conceptualization of states and reproducible alignment. However, understanding how a simple sequence of only four letters (A, T, C and G) has generated and continues to generate the great diversity of morphological,

---

Chapter written by Amir YASSIN.

physiological and ethological characters which define taxa remains a question that is still far from being resolved.

The Drosophilidae family (vinegar flies or pomace flies) is probably one of the best models to address this issue. On one hand, the family contains *Drosophila melanogaster*, the multicellular organism whose genetics and developmental biology are best known thanks to a research tradition that has been around for more than 100 years. Research on Drosophila has been rewarded with five Nobel Prizes (in 1933, 1946, 1995, 2013 and 2017), ranging from locating genes in the genome to understanding the role of these genes in development, immunity and behavior. Drosophila entered the genomic era in 2000, with the sequencing of the entire genome of *D. melanogaster* (Adams *et al.* 2000), which was followed by the sequencing of the genomes of the other 11 species in 2007 (Clark *et al.* 2007). Currently, almost 50 species have the full genome available, and for *D. melanogaster* alone, the genomes of more than 1,000 natural lineages have been sequenced (Lack *et al.* 2016)[1], providing a unique opportunity to study the molecular evolution of these 15,000 genes at different taxonomic scales.

On the other hand, the morphological and ecological diversity of the Drosophilidae family is remarkable; it is rich, with almost 5,000 described species and perhaps thousands more that remain to be discovered and described. However, the systematics of this star of genomics is still unclear, if only because of the nomenclatural problem that gives rise to the name *Drosophila melanogaster*. Indeed, the genus *Drosophila* is polyphyletic: its eight subgenera, including *Sophophora* to which *D. melanogaster* belongs, intersect with 30 other genera of Drosophilidae (Yassin 2013). Resolving this situation by raising these subgenera to the rank of genus, as is customary in current systematic practice, would change the name of *D. melanogaster*, a change that would never be accepted by the large community of Drosophila biologists. Putting these 30 genera in synonymy with *Drosophila* would have led to great taxonomic instability, due to the high incidence of synonymies of species names in the family. A debate took place in the late 2000s (Van der Linde and Yassin 2010). Although it did not solve the nomenclatural problem, it drew attention to the need to solve the systematic, and especially phylogenetic problems of the Drosophilidae (O'Grady and DeSalle 2018).

---

1 www.flypop.com.

The objective of this chapter is to show the contribution of the accumulated knowledge on genomics, genetics and developmental biology in some Drosophila species to the systematics of the family. Each of the following three sections (sections 4.2, 4.3 and 4.4) corresponds to a different taxonomic scale studied by our laboratory. First, we will deal with the genomic landscape during the early stages of speciation, taking, by way of an example, the case of the two nascent subspecies of *Drosophila yakuba*. We will then deal with the genomic landscape in the later stages of speciation, taking the evolution of a potential case of sexual color mimicry in the *Drosophila erecta* complex as an example. At these stages, questions of incongruence between the gene trees and the species trees are predominant. Finally, we will discuss phylogenetic links within the Drosophilidae family, pointing out the promises, as well as the limits, of the application of genomics to this issue. Since these three cases of subspecies, species complex and supraspecific ranks represent the three problematic situations that systematists generally face, we hope that the cases discussed below will be useful for the systematics of many other groups of organisms for which genomics tools are still under development.

4.2. Subspecies: identification of "genomic islands of divergence"?

Studies that have compared the genomes of different taxa along the speciation continuum have shown an almost-general pattern (Seehausen *et al.* 2014) (Figure 4.1). At very early stages, such as those in populations belonging to the same species, the level of divergence is often low and homogeneous throughout the genome. At very late stages, such as those between *bona fide* species, the level of divergence is high, as well as homogeneous throughout the genome. However, in the intermediate stages, the level of divergence is often heterogeneous, with some regions showing much higher divergence than the genome average. Comparing the average to an ocean, these regions of high divergence have been called "genomic islands of speciation"; they emerge from this ocean, postulating that these regions contain genes that are the target of selection (natural or sexual) that play a role in speciation (Turner *et al.* 2005). However, this has been challenged by other studies that have shown that internal genomic factors, such as variability in recombination rates linked to different structural variabilities of chromosomes – like inversions or heterochromatic content of centromeres or telomeres – could also generate such a landscape. These regions, whose role in speciation has yet to be proven, have therefore been named "genomic islands of divergence" (Wolf and Ellegren 2017).

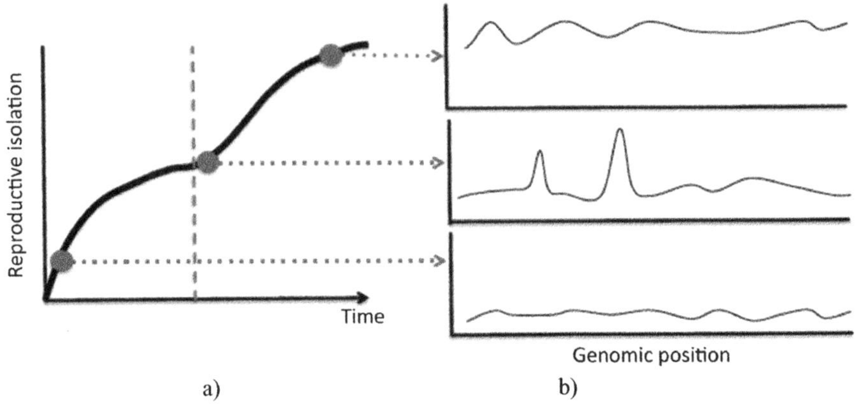

Figure 4.1. *The continuum of ecological speciation. a) Evolution of reproductive isolation over time during speciation, according to Rundle and Nosil's model (2005). b) Allelic differentiation along the genome between populations or nascent species during speciation, measured by FST. Note the formation of "genomic islands of divergence" at intermediate stages. For a color version of this figure, see www.iste.co.uk/grandcolas/systematics.zip*

The stages during which these islands are formed correspond to what has been called the "gray zone of speciation", which often presents problems of taxonomic delineation for the systematist (de Queiroz 2007). Taxa within this zone are often described as subspecies (Patten 2015). For *D. melanogaster*, the closest species is *D. simulans*, which was discovered among laboratory lineages collected in the New York area (Sturtevant 1921). Both species show clear morphological, physiological, behavioral and genetic differences, and their crossing in the laboratory is generally quite difficult (Capy and Gibert 2004). This is why, for a very long time, research on speciation in Drosophila has focused on other groups of species – in which subspecies have been identified. Take, for example, the cases of *D. pseudoobscura pseudoobscura* and *D. pseudoobscura persimilis* of the *obscura* group, or the astonishing case of the seven "races" of *D. paulistorum* of the *willistoni* group (Tsacas and Bocquet 1976). Moreover, *D. melanogaster* and *D. simulans* are cosmopolitan species, associated with human environments, which has made it difficult to study their evolutionary history.

This situation changed in the second half of the 20th century. First, a third related species, *D. yakuba*, was discovered in Côte d'Ivoire by the Swiss entomologist Hans Burla. Then, between 1970 and 2000, six other species that were endemic to tropical Africa were discovered by a group of

French researchers at the *Muséum national d'Histoire naturelle* (French National Museum of Natural History) in Paris and at the CNRS laboratory in Gif-sur-Yvette, thereby confirming the Afrotropical origin of the *melanogaster* subgroup (David *et al.* 2007). This group of researchers, which have been called The French School by drosophilists (O'Grady and DeSalle 2018), presented a unique case where systematists (L. Tsacas and M.-T. Chassagnard), ecologists (D. Lachaise), physiologists (J.R. David), cytogeneticists (F. Lemeunier) and molecular biologists (M. Solignac, M.-L. Cariou, M. Veuille, P. Capy) studied speciation in Drosophila in an integrative manner. The entire genome of the nine species of the *melanogaster* subgroup was sequenced and, indeed, the degree of genomic divergence positively correlated with the degrees of pre- and post-zygotic reproductive isolation between the nine species (Turissini *et al.* 2018). However, the nine species are quite distinct and their genomic divergences are quite large.

Perpetuating the tradition of this school, and along with J.R. David, we have carried out several missions to collect Drosophila in Madagascar and the islands of the Western Indian Ocean between 2008 and 2018, which are parts of the Afrotropical region among the least explored.

The most interesting discovery was made in 2013; it was a population of *D. yakuba* on the island of Mayotte in the Comoros archipelago (Yassin *et al.* 2016b). Morphologically, this population was indistinguishable from other populations of *D. yakuba*, which are widespread in sub-Saharan Africa, from Sierra Leone to Madagascar. However, there were two key differences. First, the Mahoran flies were only found on the fruits of noni (*Morinda citrifolia*) (Figure 4.2(a)) and never entered banana traps, the traditional method for capturing *D. yakuba* and other generalist species in the *melanogaster* subgroup in Africa. Partial precopulatory reproductive isolation was also detected in the laboratory. When Mahoran females have the choice between a male from their population or a male from the mainland, they almost always choose Mahoran males. This strong selection does not exist in the mainland females. Genomic analysis based on short autosomal introns which are less than 86 base pairs (bp) in length, and are assumed to evolve in a neutral manner (Garud *et al.* 2015), has estimated the divergence between the Mahoran population and the mainland at 29,000 years. This demographic model was then used to identify genomic regions that deviated from the genome average, and we identified 29 regions falling into regions with high recombination rates, consistent with the "genomic islands of divergence" model explained earlier (Figure 4.2(b)). For these geographic, ecological, behavioral and genomic differences, we have described the Mahoran population as a new subspecies, *D. yakuba mayottensis* (Yassin 2017). Curiously, no difference existed in the mitochondria and among 78,020,004 nuclear sites, only 36 single nucleotide polymorphisms (SNPs) distinguished *D. y. mayottensis* from the continental populations of *D. y. yakuba*.

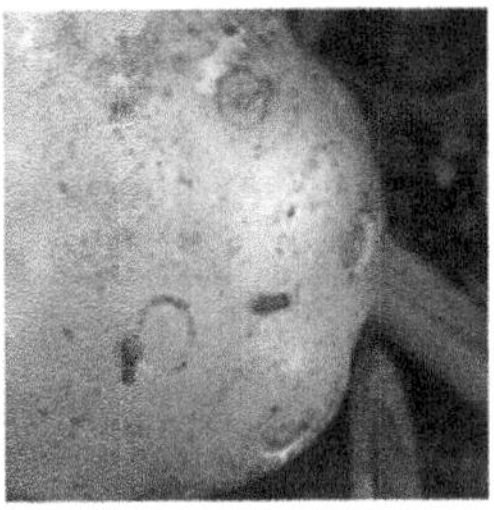

a)

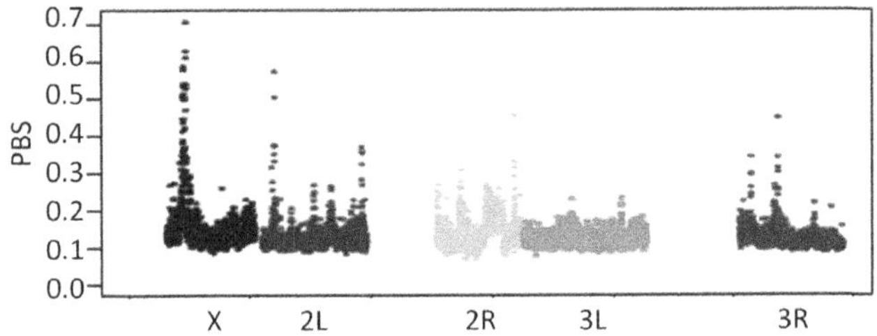

b)

Figure 4.2. *Recurrent specialization by Drosophila on toxic nono fruits (Morinda citrifolia). For a color version of this figure, see www.iste.co.uk/grandcolas/systematics.zip*

COMMENT ON FIGURE 4.2. – *a) Two Drosophila on a fallen fruit (photo by Vincent Debat). b) Genomic islands of the divergence between the specialist subspecies* D. yakuba mayottensis *and two continental populations of the generalist subspecies* D. y. yakuba, *measured by population branch statistic (PBS) (adapted from Yassin* et al. *(2016b)). Each chromosomal arm is represented by a different color. The telomeric and centromeric regions have been omitted due to their low recombination rate.*

This situation is similar to another case in the *melanogaster* subgroup. Indeed, the ripe noni fruits are toxic to most insects except for *D. sechellia*, a species that is endemic to the Seychelles, whose common ancestor with *D. yakuba* existed almost 10 million years ago (David *et al.* 2007). However, the divergence between *D. sechellia* and its generalist sister species *D. simulans* is quite large, estimated at almost 125,000 years, and their genomes are quite different (Garrigan *et al.* 2012). Moreover, crosses between these two species are difficult due to the existence of several genomic incompatibilities in their hybrids. These difficulties make the genomic mapping of the genes that potentially played a role in the speciation of *D. sechellia* and its specialization on noni difficult. Indeed, all mapping efforts to date have identified large quantitative trait loci (QTLs) comprising hundreds of genes. It is interesting to note that the majority of the genomic islands of divergence that we identified in *D. y. mayottensis* fell within these QTLs of *D. sechellia*, in a way that could not be attributed to chance, indicating that the phenotypic convergences between these two taxa may also have common genomic bases (Yassin *et al.* 2016b). In order to identify these potential bases, extensive genetic studies could be conducted under strict experimental conditions. A total of 17 phenotypes of food specialization and 20 phenotypes of reproductive isolation have been identified and are currently being analyzed in our laboratory. Once the phenomenal landscape of subspecies is well characterized, we will move on to used genetic analyses, taking advantage of the transgenesis and genomic editing tools by CRISPR\Cas9 developed in *D. yakuba* and *D. sechellia* (Stern *et al.* 2017).

4.3. Species complexes: congruence between species trees and gene trees

In the late stages of speciation, native species tend to exhibit a mosaic of traits that they share with closely related species. On a molecular level, this has often led to problems of incongruence between species trees and gene trees. In the genomic era, one of the most widely used tests for incongruence is known as the ABBA-BABA (Green *et al.* 2010). For a bifurcated tree of four species, a phylogenetically informative site with two alleles A or B could have one of three configurations: (A,(A,(B,B))), (A,(B,(B,A))) or (B,(A,(B,A))) (Figure 4.3). The first configuration is congruent with the taxa tree, and if this tree has been inferred from genomic sequences, the majority of sites on the genome should correspond to this configuration. The last two configurations are incongruent with the taxa tree and could be the result of two phenomena: multiple origins of the two similar alleles (homoplasy) or single origin of the two similar alleles not followed by phylogeny, either because of incomplete sorting of an ancestral polymorphism or because of rapid introgression between close species (hemiplasy) (Avise and Robinson 2008). Although many theoretical models have been developed to discriminate between these two phenomena or to explain the prevalence of one of them under particular conditions (Martin *et al.* 2015), the impact of these incongruities on phenotypic evolution remains to be studied.

We recently studied a potential case of sexual mimicry of coloration in the *erecta* species complex (Yassin *et al.* 2016a), a member of the *melanogaster* subgroup, which showed similarities with the configuration (A,(B,(B,A)) (Figure 4.4(a)). In the majority of species in the subgroup, males have black abdominal tips, while females have a lighter pigmentation. In contrast, in the *D. erecta* species alone, females show a clear Mendelian dimorphism with a black "andromorphic" shape that resembles males, and a light "gymnomorphic" shape that is completely devoid of any pigment (Figure 4.4(b)). This phenomenon has been widely studied in dragonflies, where andromorphic shapes are considered as an adaptive frequency-dependent strategy to reduce the costs of sexual harassment (Takahashi *et al.* 2014). However, the lack of advanced genetic tools in dragonflies prevents the identification of the underlying locus. In *D. orena*, the sister species of *D. erecta*, a single light morph is present. The most parsimonious scenario would be that from an ancestor with blackish females like *D. melanogaster*, a light morph has evolved in the ancestor of the *erecta* complex, then within *D. erecta*, a new allele, a reversion to the ancestral black state has evolved.

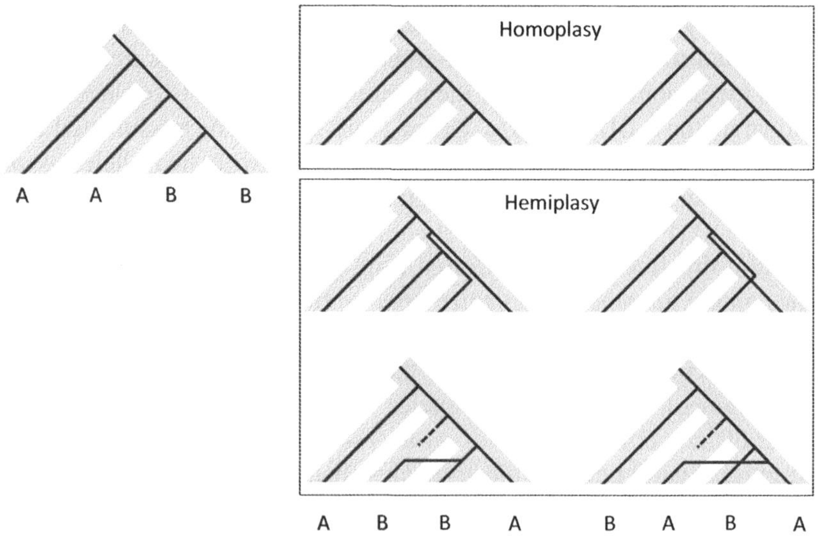

Figure 4.3. *ABBA-BABA genomic test, demonstrating different possible scenarios of incongruence between bi-allelic genomic sites (alleles colored in blue and red) and the tree of species (in gray). For a color version of this figure, see www.iste.co.uk/grandcolas/systematics.zip*

In order to test this hypothesis, we started by mapping this character in *D. erecta*. First, we selected homozygous lineages for the black allele and others for the light allele; as the black allele was dominant in the heterozygote, we introduced it by backcross in a clear genetic background for seven generations. Then, we sequenced the light and black parental lineages and the black introgressed flies. Genomic analysis then showed that a small region of almost 1 kilobase (kb) upstream of the *tan* gene for melanin synthesis, was enriched with alleles from the black parent in the introgressed flies (Figure 4.4(c)). In order to test the potential role of this region in dimorphism, we used transgenic fluorescent markers (Figure 4.4(d)). This approach first consists of amplifying the black and light haplotypes by PCR (Polymerase Chain Reaction) and then ligating them separately with a GFP (Green Fluorescent Protein) from the bioluminescent jellyfish *Aequorea victoria*. The construct is then ligated to a plasmid, which is a non-chromosomal DNA present in bacterial cells, probably of viral origin. The plasmid is then cloned into a bacterium in order to amplify it and, after being isolated and purified, it is injected into an egg from a transgenic *D. melanogaster* lineage. The genome of this lineage has a particular site; thanks to transposons (mobile genetic elements) present on the plasmid, the constructions containing the two alleles are inserted into the same region of the genome. In this experiment, we found that the region from the black morph of *D. erecta* expressed GFP in the tip of the abdomen of *D. melanogaster* females, while the region from the light morph did not, thus confirming the role of this region in the color dimorphism of *D. erecta* females.

We then sequenced this regulatory region of the *tan* gene in *D. orena*, the sister species of *D. erecta*, whose females are light. To our surprise, the sequence of *D. orena* was closer to the black haplotype of *D. erecta* than to the light one (Figure 4.4(a)). Indeed, *D. orena* almost had the same sequences as the black *D. erecta*, except for a single insertion of about 50 bp that was located right in the middle of the most conserved region of the black haplotype, and which more likely serves as a binding site for the transcription factor controlling the expression of the *tan* gene (Figure 4.4(c)). This suggests that, although the dimorphism in *D. erecta* is likely ancestral, the light allele in *D. orena* is more likely derived from the ancestral black allele, indicating a new, less morphologically parsimonious molecular topology that generates the pattern (A,(B,(B,A)). This indicates the contribution of genetic studies to the understanding of the mechanisms of incongruence between species and trait trees in the late stages of speciation.

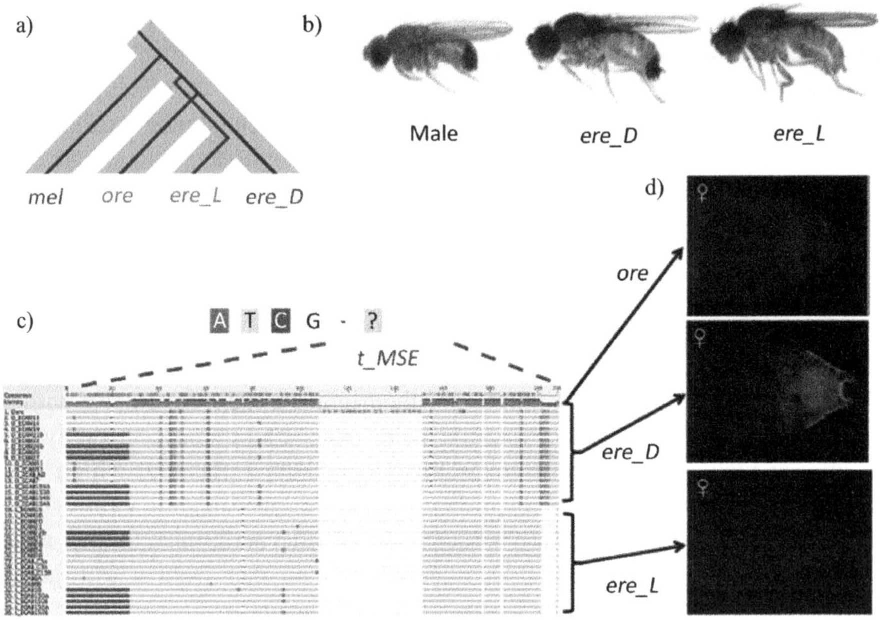

Figure 4.4. *Complex history of the evolution of the allele of the abdominal pigmentation of females of the erecta complex. For a color version of this figure, see www.iste.co.uk/grandcolas/systematics.zip*

COMMENT ON FIGURE 4.4. – *a) Tree of species (in gray) against the tree of black (in red) and light (in blue) alleles: mel* = melanogaster, *ore* = orena, *ere* = erecta. *b) Black morph (ere_D) and light morph (ere_L) of* D. erecta *females compared to males (photo by Michel Veuille). c) Haplotypic diversity of black and light alleles in* D. erecta. *Note the similarity between the light haplotype of* D. *orena and the black haplotypes of* D. erecta, *as well as the insertion of about 50 bp in the middle of the cis-regulatory region of the* tan *gene (t_MSE). d) Functional validation by transgenesis of the GFP markers of the three haplotypes (from Yassin et al. 2016a).*

4.4. Supraspecific ranks: phylogeny, genome and morphome

The Drosophilidae family comprises about 5,000 species that are often classified under seven supraspecific ranks: complex, subgroup, group, subgenus, genus, tribe and subfamily. Clarifying the phylogenetic links between these categories is essential in order to infer a complete picture of the evolution of this family. We have proposed a phylogenetic classification of the family, based on the molecular sequences of eight genes (seven nuclei and one mitochondrial: 8,248 bp) for several species belonging to 33 genera and several groups of species (Yassin 2013). For the categories for which molecular sequences were not available, we searched for the morphological traits most congruent with the molecular tree and used them to graft these categories on the tree. We have called this approach "morphological grafting" (Yassin *et al.* 2008), and although it still lacks a robust statistical framework (O'Grady and DeSalle 2018), it has had the advantage of presenting an exhaustive classification for the family and of formulating several phylogenetic hypotheses that are awaiting validation. We can envision the future of Drosophilidae phylogenetics going in two complementary directions.

On one hand, more and more species genomes will be sequenced. Of course, we must initially expect a certain bias in favor of "generalist" species that are bred in laboratories, or that are currently found in large stock centers which are of major interest to geneticists, such as the National Drosophila Species Stock Center in New York, or the EHIME-Fly in Japan. Ultimately, these efforts will produce a large number of "standard" genes for the phylogenetic placement of hundreds of species, even those that are not bred in the laboratory and whose genetics remain completely unknown. These include those associated with ground crabs or the spits of cercopids, those that eat whiteflies, mealybugs, simuliids, chironimids, dragonfly eggs, spider cocoons or frog eggs, and those that parasitize wasps (for a review, see Ashburner (1981)).

However, even if these approaches will molecularly place several species on the Drosophilidae tree, many others only remain known by a few rare specimens, or even fossils, and an exhaustive classification will require the use of morphological characters.

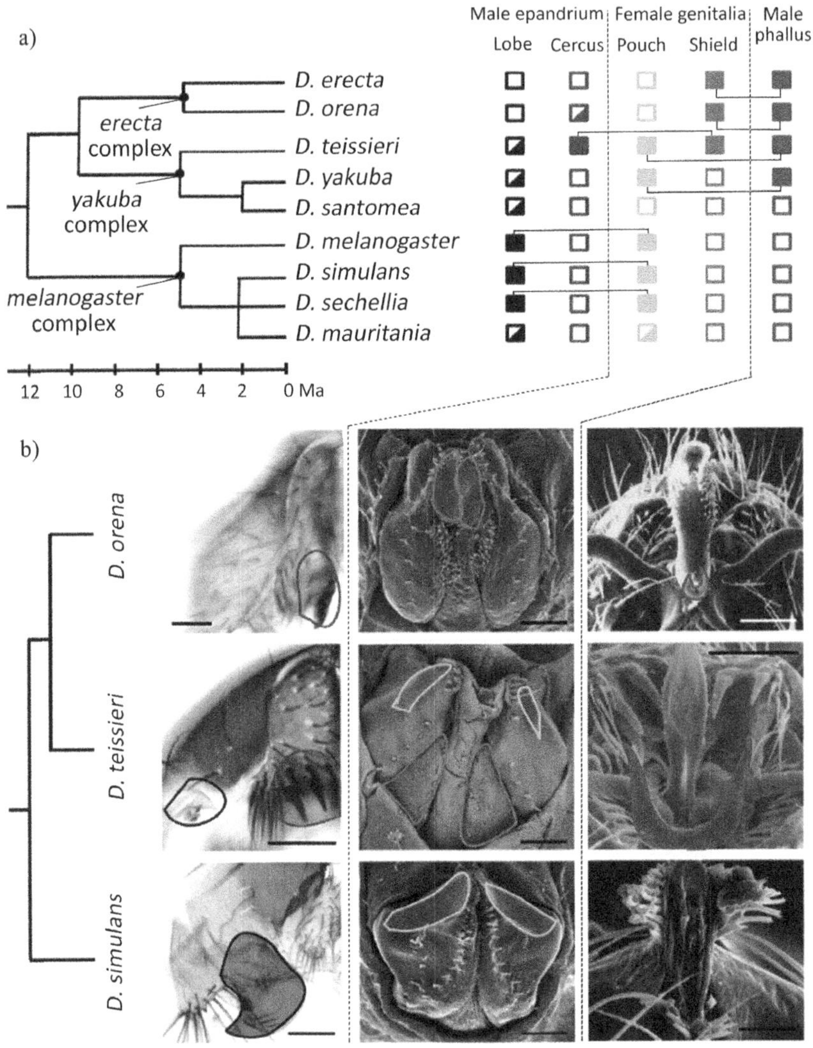

Figure 4.5. *Coevolution between male (blue, green and red) and female (yellow and orange) genitalia in the nine species of the Drosophila melanogaster subgroup. For a color version of this figure, see www.iste.co.uk/grandcolas/systematics.zip*

COMMENT ON FIGURE 4.5. – *a) Phylogenetic comparison of the anatomical interactions of homologous structures during mating. Note that homologous structures of one sex do not interact with the same structures of the opposite sex in different species. b) Illustration of genitalia in three species. The vestigial structures are surrounded, but not filled (e.g. the lobe of the epandrium in* D. teissieri *(blue) and the spines of the cerci in* D. orena *(green)). Homologous structures in females do not interact with homologous structures in males. For example, ovipositor pouches in* D. teissieri *and* D. simulans *(yellow) interact with the lobe of the epandrium in* D. simulans *(blue) and phallic spines in* D. teissieri *(red). Similarly, the vaginal shields in* D. teissieri *and* D. orena *(orange) interact with the spines of the cerci in* D. teissieri *(green) and the phallus spines in* D. orena *(red).*

Thus, these genomic efforts will have to be complemented by morphological analyses, so that a reference "morpheme" can also be reconstructed for Drosophila. To this end, we started by analyzing the taxonomic literature in order to document the morphological diversity described in several species. An analysis of two important monographs on the juvenile and adult stages of the 56 species representing the four major clades of the family allowed the conceptualization of 490 morphological characters (Al Sayad and Yassin 2019). It has also shown, by studying the evolution of these characters on the molecular tree, that two-thirds of the morphological changes are homoplastic; in other words, they lead to the apparition of the same character independently in distant species, thus showing the limitations of morphology alone in phylogenetics. We have also participated, together with systematists and geneticists, in the production of an ontology of terms related to the morphology of male genitalia in *D. melanogaster* (Rice *et al.* 2019), structures that evolve most rapidly between species (Yassin and Orgogozo 2013) (Figure 4.5) and that probably contribute to mechanical reproductive isolation during speciation (Yassin 2016; Yassin and David 2016). Ontological revisions of other organs are also needed; in addition, more systematized, and preferably more computerized, documentation of the majority of Drosophilidae species should be pursued, in order to better understand how the diversity in the genomes of the species to be sequenced has potentially contributed to the origin and evolution of the diversity of characters, as well as morphological, physiological and ethological adaptations of this large family.

4.5. Conclusion

The genomic revolution has revolutionized and continues to revolutionize the life sciences, and systematics is no exception. It represents a very promising common factor for the integration of different disciplines of biology into a single framework. While the genomic era is characterized by the sequencing of more and more genomes of different species combined with efforts to assemble, annotate and align genomes, the post-genomic era opens a door that has previously only been open to a few organisms on the underlying mechanisms of origin and maintenance of diversity. Drosophila, a model organism par excellence for genetic research, with contributions to medicine and agronomy, has always contributed to the development of systematics. It has been suggested that Hennig, who was initially a dipterologist, was greatly influenced by the meticulous description of the origin and changes of traits in early efforts to establish a "*satisfactory arrangement of [Drosophila] species into some scheme of classification that can be taken as indicating their degree of genetic relationship*" by Sturtevant (O'Grady and DeSalle 2018). Dobzhansky's meticulous studies of reproductive isolation in *D. pseudoobscura* helped to formulate the "biological concept of the species", one of the first concepts that would greatly influence systematists' approaches to species delineation (Coyne and Orr 2004). In the post-genomic era, new approaches concerning the mechanisms of genome evolution, ecological speciation, morphological convergences or trait conceptualization could be developed from Drosophila studies, which have approaches that definitely go beyond this model. Understanding how the great diversity of life is preserved and mostly generated by DNA sequences adds greatness to Darwin's (1859) famous vision of life as "*from so simple a beginning, endless forms most beautiful have been, and are being, evolved*".

4.6. Acknowledgments

The author would like to thank the contributors who worked on different aspects of these projects: Jean R. David, John E. Pool, Héloïse Bastide, Vincent Debat, Nelly Gidaszweski, Arnaud Suwalski, Henry Chung, Michel Veuille, Virginie Courtier and Sinan Al Sayad. Part of the research is funded by ANR-18-CE02-0008.

4.7. References

Adams, M.D., Celniker, S.E., Holt, R.A., Evans, C.A., Gocayne, J.D., Amanatides, P.G. *et al.* (2000). The genome sequence of *Drosophila melanogaster* [Online]. *Science*, 287(5461), 2185–2195. Available: https://doi.org/10.1126/science.287.5461.2185.

Al Sayad, S. and Yassin, A. (2019). Quantifying the extent of morphological homoplasy: A phylogenetic analysis of 490 characters in Drosophila [Online]. *Evolution Letters*, 3(3), 286–298. Available: https://doi.org/10.1002/evl3.115.

Ashburner, M. (1981). Entomophagous and other bizarre Drosophilidae. In *The Genetics and Biology of Drosophila*, Ashburner, M., Carson, H.L., and Thompson, J.N. (eds). Academic Press, Cambridge, 395–429.

Avise, J.C. and Robinson, T.J. (2008). Hemiplasy: A new term in the lexicon of phylogenetics [Online]. *Systematic Biology*, 57(3), 503–507. Available: https://doi.org/10.1080/10635150802164587.

Berlocher, S.H. (1984). Insect molecular systematics [Online]. *Annual Review of Entomology*, 29(1), 403–433. Available: https://doi.org/10.1146/annurev.en.29.010184.002155.

Capy, P. and Gibert, P. (2004). *Drosophila melanogaster, Drosophila simulans*: So similar yet so different. In *Drosophila Melanogaster, Drosophila Simulans: So Similar, So Different*, Capy, P., Gibert, P., and Boussy, I. (eds). Springer, Netherlands, Dordrecht, 5–16.

Clark, A.G., Eisen, M.B., Smith, D.R., Bergman, C.M., Oliver, B., Markow, T.A. *et al.* (2007). Evolution of genes and genomes on the Drosophila phylogeny [Online]. *Nature*, 450(7167), 203–218. Available: https://doi.org/10.1038/nature06341.

Coyne, J.A. and Orr, H.A. (2004). *Speciation*. Sinauer Associates, Cary, North Carolina.

Darwin, C. (1859). *On the Origin of Species by Means of Natural Selection, Or, the Preservation of Favoured Races in the Struggle for Life*. J. Murray, London.

David, J.R., Lemeunier, F., Tsacas, L., and Yassin, A. (2007). The historical discovery of the nine species in the *Drosophila melanogaster* species subgroup. *Genetics*, 177(4), 1969–1973.

Garrigan, D., Kingan, S.B., Geneva, A.J., Andolfatto, P., Clark, A.G., Thornton, K.R., and Presgraves, D.C. (2012). Genome sequencing reveals complex speciation in the *Drosophila simulans* clade [Online]. *Genome Research*, 22(8), 1499–1511. Available: https://doi.org/10.1101/gr.130922.111.

Garud, N.R., Messer, P.W., Buzbas, E.O., and Petrov, D.A. (2015). Recent selective sweeps in North American *Drosophila melanogaster* show signatures of soft sweeps [Online]. *PLOS Genetics*, 11(2), e1005004. Available: https://doi.org/10.1371/journal.pgen.1005004.

Green, R.E., Krause, J., Briggs, A.W., Maricic, T., Stenzel, U., Kircher, M. *et al.* (2010). A draft sequence of the Neandertal Genome [Online]. *Science*, 328(5979), 710–722. Available: https://doi.org/10.1126/science.1188021.

Lack, J.B., Lange, J.D., Tang, A.D., Corbett-Detig, R.B., and Pool, J.E. (2016). A thousand fly genomes: An expanded *Drosophila* genome Nexus [Online]. *Molecular Biology and Evolution*, 33(12), 3308–3313. Available: https://doi.org/10.1093/molbev/msw195.

Martin, S.H., Davey, J.W., and Jiggins, C.D. (2015). Evaluating the use of ABBA-BABA statistics to locate introgressed loci [Online]. *Molecular Biology and Evolution*, 32(1), 244–257. Available: https://doi.org/10.1093/molbev/msu269.

O'Grady, P.M. and DeSalle, R. (2018). Phylogeny of the genus *Drosophila* [Online]. *Genetics*, 209(1), 1–25. Available: https://doi.org/10.1534/genetics.117.300583.

Patten, M.A. (2015). Subspecies and the philosophy of science: A philosophy of subspecies [Online]. *The Auk*, 132(2), 481–485. Available: https://doi.org/10.1642/AUK-15-1.1.

de Queiroz, K. (2007). Species concepts and species delimitation [Online]. *Systematic Biology*, 56(6), 879–886. Available: https://doi.org/10.1080/10635150701701083.

Rice, G., David, J.R., Kamimura, Y., Masly, J.P., Mcgregor, A.P., Nagy, O. *et al.* (2019). A standardized nomenclature and atlas of the male terminalia of *Drosophila melanogaster* [Online]. *Fly*, 13(1-4), 51–64. Available: https://doi.org/10.1080/19336934.2019.1653733.

Rundle, H.D. and Nosil, P. (2005). Ecological speciation [Online]. *Ecology Letters*, 8(3), 336–352. Available: https://doi.org/10.1111/j.1461-0248.2004.00715.x.

Seehausen, O., Butlin, R.K., Keller, I., Wagner, C.E., Boughman, J.W., Hohenlohe, P.A. *et al.* (2014). Genomics and the origin of species [Online]. *Nature Reviews Genetics*, 15(3), 176–192. Available: https://doi.org/10.1038/nrg3644.

Stern, D.L., Crocker, J., Ding, Y., Frankel, N., Kappes, G., Kim, E. *et al.* (2017). Genetic and transgenic reagents for *Drosophila simulans*, *D. mauritiana*, *D. yakuba*, *D. santomea* and *D. virilis* [Online]. *G3: Genes, Genomes, Genetics*, 7(4), 1339–1347. Available: https://doi.org/10.1534/g3.116.038885.

Sturtevant, A.H. (1921). *The North American Species of Drosophila*. Carnegie Institute of Washington Publication, Washington.

Takahashi, Y., Kagawa, K., Svensson, E.I., and Kawata, M. (2014). Evolution of increased phenotypic diversity enhances population performance by reducing sexual harassment in damselflies [Online]. *Nature Communications*, 5(1), 1–7. Available: https://doi.org/10.1038/ncomms5468.

Tsacas, L. and Bocquet, C. (1976). L'espèce chez les Drosophilidae. In *Les problèmes de l'espèce dans le règne animal*, Bocquet, C., Génermont, J., and Lamotte, M. (eds). Société zoologique de France, Paris, 203–247.

Turissini, D.A., McGirr, J.A., Patel, S.S., David, J.R., and Matute, D.R. (2018). The rate of evolution of postmating-prezygotic reproductive isolation in Drosophila [Online]. *Molecular Biology and Evolution*, 35(2), 312–334. Available: https://doi.org/10.1093/molbev/msx271.

Turner, T.L., Hahn, M.W., and Nuzhdin, S.V. (2005). Genomic Islands of speciation in *Anopheles gambiae* [Online]. *PLOS Biology*, 3(9), e285. Available: https://doi.org/10.1371/journal.pbio.0030285.

Van der Linde, K. and Yassin, A. (2010). A fly by any other name. *New Scientist*, 206, 24–25.

Wolf, J.B.W. and Ellegren, H. (2017). Making sense of genomic islands of differentiation in light of speciation [Online]. *Nature Reviews Genetics*, 18(2), 87–100. Available: https://doi.org/10.1038/nrg.2016.133.

Yassin, A. (2013). Phylogenetic classification of the Drosophilidae Rondani (Diptera): The role of morphology in the postgenomic era. *Systematic Entomology*, 38(2), 349–364.

Yassin, A. (2016). Unresolved questions in genitalia coevolution: Bridging taxonomy, speciation, and developmental genetics [Online]. *Organisms Diversity & Evolution*, 16(4), 681–688. Available: https://doi.org/10.1007/s13127-016-0286-2.

Yassin, A. (2017). *Drosophila yakuba mayottensis*, a new model for the study of incipient ecological speciation [Online]. *Fly*, 11(1), 37–45. Available: https://doi.org/10.1080/19336934.2016.1221550.

Yassin, A. and David, J.R. (2016). Within-species reproductive costs affect the asymmetry of satyrization in Drosophila. *Journal of Evolutionary Biology*, 29(2), 455–460.

Yassin, A. and Orgogozo, V. (2013). Coevolution between male and female genitalia in the *Drosophila melanogaster* species subgroup. *PLOS One*, 8(2), e57158.

Yassin, A., Araripe, L.O., Capy, P., Da Lage, J.-L., Klaczko, L.B., Maisonhaute, C. *et al.* (2008). Grafting the molecular phylogenetic tree with morphological branches to reconstruct the evolutionary history of the genus *Zaprionus* (Diptera: Drosophilidae). *Molecular Phylogenetics and Evolution*, 47(3), 903–915.

Yassin, A., Bastide, H., Chung, H., Veuille, M., David, J.R. *et al.* (2016a). Ancient balancing selection at tan underlies female colour dimorphism in *Drosophila erecta* [Online]. *Nature Communications*, 7, 10400. Available: https://doi.org/10.1038/ncomms10400.

Yassin, A., Debat, V., Bastide, H., Gidaszewski, N., David, J.R., and Pool, J.E. (2016b). Recurrent specialization on a toxic fruit in an island Drosophila population [Online]. *Proceedings of the National Academy of Sciences*, 113(17), 4771–4776. Available: https://doi.org/10.1073/pnas.1522559113.

5

Dealing with Multiple Environments: The Challenges of the Trypanosome Lifecycle

5.1. Human African trypanosomiasis, the disease

African trypanosomes (*Trypanosoma* genus) are unicellular parasites responsible for infectious diseases that affect humans and animals. In humans, this extracellular protist causes Human African trypanosomiasis (HAT), also known as "sleeping sickness", an under-studied tropical disease that is restricted to populations in rural sub-Saharan Africa, where the tsetse fly vector (*Glossina* genus) transmits the parasite. Human African trypanosomiasis is characterized by a markedly focal distribution that depends on complex interactions between parasite, vector, host and the environment, important aspects that will be presented here.

Human African trypanosomiasis is caused by two subspecies: *Trypanosoma brucei gambiense*, that leads to a slow progression and is endemic in western and central Africa, accounting for 98% of the cases; and *T. b. rhodesiense*, a zoonotic pathogen that progresses faster and is present in eastern and southern Africa. Although devastating human epidemics have occurred over the last century, sustained and coordinated control efforts over the past 20 years have resulted in an overall 95% case reduction (Franco *et al.* 2020) and less than 1,000 cases were reported in 2018, an unprecedented accomplishment (WHO website: www.who.int/trypanosomiasis). Given the fact that sleeping sickness occurs in remote, difficult-to-access rural areas, it is likely that more cases remain undetected.

Chapter written by Estefanía CALVO ALVAREZ and Philippe BASTIN.

The clinical presentation of HAT depends on the subspecies involved, the host immune response and the stage of the disease. Both forms of the disease are potentially fatal if left untreated. However, for *gambiense* HAT, parasite clearance and human trypanotolerance have been recognized in the absence of treatment (Jamonneau *et al.* 2012). *T. b. gambiense* HAT is characterized by an early hemo-lymphatic stage (stage 1) which can occur one to two weeks after the initial infection and can persist for several months to years (Jamonneau *et al.* 2012); it is then followed by a meningo-encephalitic late stage (stage 2) where parasites invade the central nervous system (CNS). During this stage, neuropsychiatric disorders accompany first-stage features. Indeed, the name "sleeping sickness" derives from the characteristic sleep disorder found at stage 2, which is accompanied by manic episodes, confusion and dementia (Büscher *et al.* 2017).

5.2. Cell biology of *Trypanosoma brucei*

Trypanosoma brucei is a eukaryotic flagellated parasite that alternates extracellularly between two different hosts – the mammalian host and the arthropod vector – where it encounters very distinct microenvironments to which it must adapt in order to complete its complex developmental cycle and, eventually, be transmitted.

During evolution, trypanosomes diverged from the main eukaryotic lineage, but the assumption that they represent an ancestral state is misleading, as they co-evolved in parallel with their more speciated animal hosts (Dacks *et al.* 2008). Indeed, recent revisions of eukaryotic systematics and taxonomy have established the division of eukaryotes into six major supergroups based on a combination of morphological and molecular analyses (Simpson and Roger 2004; Adl *et al.* 2012). Among them, the Excavata group contains many parasitic lineages, including the genus *Trypanosoma*. *T. brucei* species belong to the sub-kingdom Protozoa, the order Kinetoplastida and the family Trypanosomatidae (Adl *et al.* 2012). This new eukaryotic phylogeny emphasizes the fact that trypanosomes are not primitive, but, in contrast, highly evolved and specialized for their particular environments, which makes them valuable models for studying different eukaryotic processes.

Trypanosomes are diploid organisms. The genetic information of *T. brucei* consists of ~35 Mb (per haploid genome), arranged in 11 pairs of megabase chromosomes and enclosed in a nucleus with a diameter of 2–2.5 μm (Figure 5.1) (Ersfeld 2011). The total nuclear DNA content also includes intermediate chromosomes and mini-chromosomes that account for ~20% of the total genome size (Berriman *et al.* 2005), and whose role is exclusively associated with the process of antigenic variation (see below) (Robinson *et al.* 1999; Morrison *et al.* 2005; Hertz-Fowler *et al.* 2008).

An important and identifiable morphological feature of *T. brucei* is the presence of a single mitochondrion whose massive DNA, called the kinetoplast (kDNA) (Figure 5.1), is a diagnostic feature of the Kinetoplastida order. It is an extraordinary network structure of two topologically interlinked types of DNA rings: a few dozen ~23 kb maxicircles (homologs of the mitochondrial genome of other eukaryotes) and thousands of ~1 kb minicircles (Vickerman 1965; Jensen and Englund 2012; Lukeš *et al.* 2018). Duplication and segregation of this single mitochondrion relies on the basal body (the functional equivalent of the centriole) via a dedicated tripartite anchoring structure (Robinson Gull 1991; Ogbadoyi *et al.* 2003). The kinetoplast encodes protein components of the mitochondrial respiratory chain and associated proteins that participate in DNA compaction and replication (Simpson 1987). Kinetoplastid mitochondrial gene expression involves a cooperation between maxicircle and minicircle transcripts, since maxicircle transcripts require editing to form a functional and translatable mRNA (Benne *et al.* 1986). Editing specificity is controlled by minicircle-encoded guide RNAs (gRNA), following by the addition or deletion of uridylate residues at internal sites within the maxicircle transcripts, which, in turn, explains the large reservoir of minicircle sequences (Hong and Simpson 2003; Cooper *et al.* 2019).

Eukaryotic cells adapt to specialized functions or to changing environments through differentiation. In particular, microbial eukaryotes exhibit programmed developmental changes to maximize their proliferation, transmission and survival. The major changes that *T. brucei* parasites undergo during the course of their lifecycle are mediated by regulated changes in gene expression. Intriguingly, *T. brucei* displays atypical nuclear gene expression regulation compared to other eukaryotic organisms. The genome is organized into large polycistronic gene clusters, that is, tens-to-hundreds of protein-coding genes arranged sequentially on the same strand of DNA. The genes within polycistronic units are not functionally related but are nevertheless co-transcribed by RNA polymerase II and subsequently processed into individual and mature mRNAs by trans-splicing (the addition of a 39-nucleotide long spliced leader at the 5' end) and polyadenylation

(Michaeli 2011). Regulation is mostly controlled by the 3' untranslated region of each gene (Clayton 2019). Several developmental programs are coordinated by RNA-binding proteins that act as trans-acting factors in the control of developmental transitions in the absence of canonical transcription factors (Clayton 2013). To date, three major RNA-binding proteins have been identified (Subota *et al.* 2011; Kolev *et al.* 2012; Mugo and Clayton 2017) and several proteins containing zinc finger motifs could also modulate differentiation (Hendriks *et al.* 2001; Hendriks and Matthews 2005; Benz *et al.* 2011; Walrad *et al.* 2012).

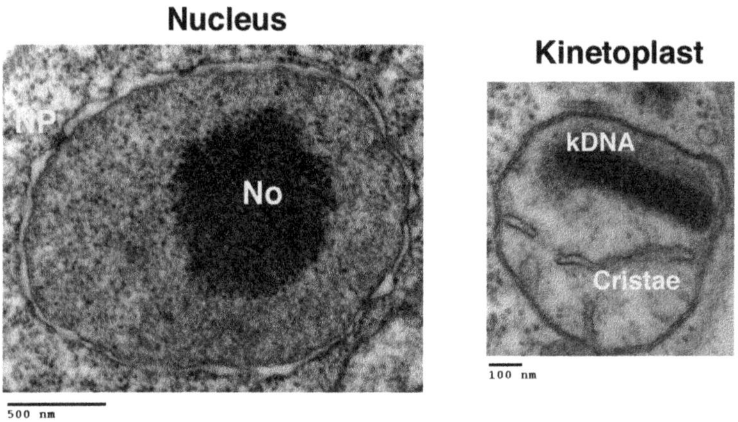

Figure 5.1. *Electron microscopy images of the nucleus (left) and kinetoplast (right) of* T. brucei *(procyclic stage). The nucleolus (No), the double membrane and some nuclear pores (NP) are visible. Kinetoplast DNA appears as a large electron-dense mass and fibers pointing towards the membrane are indicated. Typical mitochondrial cristae are visible (Toutirais and Bastin, unpublished data). For a color version of this figure, see www.iste.co.uk/grandcolas/systematics.zip*

In unicellular organisms, morphological transitions have been proposed to represent adaptations to specialized environments (Fulton 1977). Over a century ago, Plimmer and Bradford provided the first scientific description of the morphology of an "adult form of *Trypanosoma*" seen in a drop of blood of an infected dog: "… appears to consist of a uniform, homogeneous mass of protoplasm, of worm-like form, with at one end a thick, stiff extremity, and at the other a long, wavy flagellum" (Hecker *et al.* 1972). The anterior–posterior axis is defined by the single flagellum laterally attached to the cell body and emerging from the cell posterior. The flagellum mostly beats from tip to base, dragging the cell forward (Walker 1961), but can also exhibit large base to tip waves, leading to reorientation (Branche *et al.* 2006; Heddergott *et al.* 2012). *T. brucei* subspecies are morphologically indistinguishable: in the mammalian host, trypanosomes are most often observed as spindle-shaped cells that are 20–30 μm long (approximately three times the diameter of a human erythrocyte) and 2–5 μm wide, and are characterized by their wriggling movement. A shorter trypanosome form is also observed: these cells are metabolically preadapted to survive in the alimentary tract of the tsetse fly. Strikingly, while in the tsetse vector, at least seven distinct morphological forms which range from 35 to 4 μm in length have been identified in different tsetse organs (Rotureau *et al.* 2011).

These spectacular transformations in cell size and shape are accompanied by specific repositioning of organelles among parasite stages, which are presumably required for successful completion of their developmental program. The differentiated cell types of the trypanosome lifecycle are defined morphologically by the relative positions of the nucleus and the kinetoplast. This relative configuration defines the two main parasitic morphotypes (Hoare and Wallace 1966): trypomastigote, with the kinetoplast localizing between the nucleus and the posterior end of the cell; and the epimastigote arrangement, with the kinetoplast between the nucleus and the anterior end, due to the posterior migration of the nucleus to the other side of the kinetoplast (Sharma *et al.* 2008).

The shape of the cell is defined by a peripheral corset of microtubules that are cross-linked to each other and to the plasma membrane (Sherwin and Gull 1989). This corset is highly resistant to damage and might indeed be a protective adaptation for the parasite, helping it to survive within very different environments, in terms of chemical or physical properties. Newly assembled cytoskeletal microtubules are added to the posterior end of the cell so that the microtubule corset is not uniform (Sherwin *et al.* 1987).

The flagellum is a central feature of *T. brucei* cell biology and lifecycle with inherent implications for parasite adaptation. All stages of *T. brucei* possess a single flagellum, a cylindrical structure surrounded by a membrane and composed of a canonical 9+2 axoneme microtubules, the cytoskeletal core of the flagellum (Ralston *et al.* 2009). The flagellum is anchored to the cell body with the exception of its distal end and emerges from an invagination of the plasma membrane called the flagellar pocket. This is positioned in a gap in the sub-pellicular corset and is dedicated to endo- and exocytosis, acting as a major gate for host–parasite interactions (Webster and Russell 1993; Field and Carrington 2009). The flagellum of *T. brucei* is a multifaceted organelle that is crucial for key parasite functions: it is essential for parasite viability and morphogenesis (Kohl *et al.* 2003), and is responsible for parasite motility (Heddergott *et al.* 2012; Rotureau *et al.* 2014; Schuster *et al.* 2017). In the context of morphogenesis, cell division and inheritance of essential cellular structures rely on the flagellum that defines morphogenetic axes within the cell (Vaughan 2010). In fact, flagellum length controls elongation of an internal structure called the flagellum attachment zone (FAZ) that mediates lateral flagellum attachment to the plasma membrane and is probably the driver of cytokinesis (Robinson *et al.* 1995; Kohl *et al.* 2003; Vaughan *et al.* 2008; Zhou *et al.* 2011). Finally, the flagellum could act as a sensory platform and interface for host–parasite interactions, equivalent to the sensory functions reported for cilia and flagella in other eukaryotes (Singla and Reiter 2006).

5.3. Survival and maturation of *T. brucei* in the tsetse vector

Successful parasite transmission is the result of a crosstalk between trypanosomes and their tsetse fly vector. Biting flies of the genus *Glossina* are endemic to sub-Saharan Africa and are commonly known as tsetse, with tsetse meaning "fly" in the Tswana language of southern Africa (Welburn *et al.* 2016). Unlike other dipteran vectors, both sexes are obligatory hematophagous (blood feeders) and therefore both can transmit trypanosomes (Krenn and Aspöck 2012). Tsetse flies are viviparous, and the female deposits a fully developed larva that burrows into the soil, pupates, and emerges as an adult fly a month later. The 31 tsetse species and subspecies are classified as forest, riverine, or savanna, according to morphological differences and habitat preference (Cecchi *et al.* 2008).

Since they transmit trypanosomes via the saliva during fly feeding, all pathogenic African trypanosomes are called Salivarian. The plasticity of these parasites to adapt to new environments becomes evident as soon as the transmissible "stumpy" parasites present in the blood are ingested during the tsetse's blood meal on an infected mammal. It has been accepted that only the stumpy stage is pre-adapted for survival in the tsetse's midgut and thus, the only one capable of being transmitted (Rico *et al.* 2013).

Once in the fly, parasite development occurs over the course of 20–30 days (Aksoy *et al.* 2003), when different steps of parasite proliferation, migration and differentiation are accompanied by subsequent metabolic, cell surface protein remodeling and changes in morphology, flagellar length, and the relative positions of the parasite nucleus and kinetoplast (Hoare and Wallace 1966; Sharma *et al.* 2008). These developmental transitions are highly temporally ordered, and coincide with the specific fly tissues/organs where the main transformations and adaptations take place (Figure 5.2).

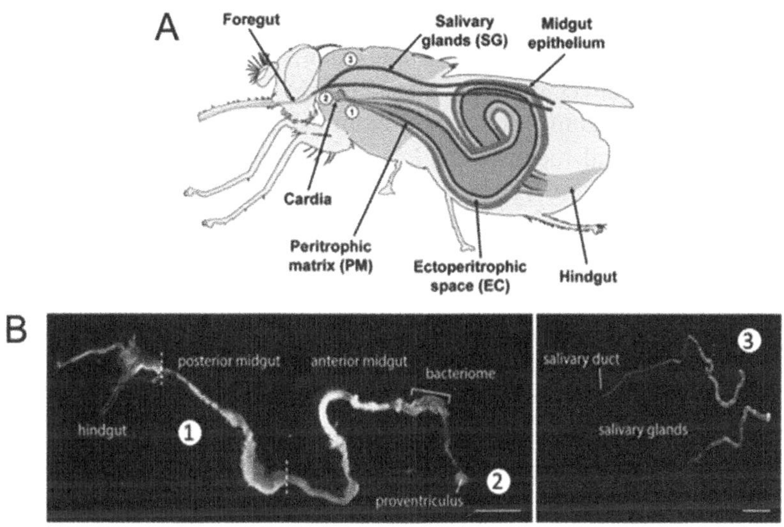

Figure 5.2. *Representation of a Glossina morsitans morsitans tsetse fly. A) Drawing of the different organs and compartments of the alimentary tract. B) Isolation of an intact alimentary tract, highlighting the midgut (1), divided into a posterior part, where the blood meal is digested, and the anterior part, which includes the bacteriome and ends in the proventriculus or cardia (2). Scale bar: 3 mm. The two salivary glands with the thinner salivary ducts are shown (3). Scale bar: 1 mm (adapted from Aksoy (2019) and Schuster* et al. *(2017)). For a color version of this figure, see www.iste.co.uk/grandcolas/systematics.zip*

In the tsetse's posterior midgut lumen, bloodstream trypanosomes differentiate into procyclic trypomastigotes, an event that occurs during the first 24 hours (Figure 5.3) (MacGregor and Matthews 2010; Savage *et al.* 2012). Freshly differentiated procyclic parasites multiply in the posterior midgut of the fly. A major and striking modification is the replacement of the surface coat consisting of variant surface glycoprotein (VSG) present in bloodstream trypanosomes by procyclins, a family of proteins characterized by multiple repetitions of glutamates and prolines (EP procyclin), or of a pentapeptide composed of glycine, proline, two glutamates and threonine

(GPEET procyclin) (Mowatt and Clayton 1987; Roditi *et al.* 1987). Both VSG and procyclin proteins are anchored to the membrane via a glycosylphosphatidylinositol-anchor and the replacement takes place via the action of a membrane-anchored metalloprotease (MSP B) that cleaves VSG from the surface (Gruszynski *et al.* 2006; Grandgenett *et al.* 2007). The insect-specific EP procyclin coat is gained after 2 hours, overlapping with the loss of the VSG coat, which occurs 4–5 hours after bloodstream parasites enter the tsetse fly midgut (Matthews 1999). At this early point, most tsetse flies can efficiently clear the infection (Sharma *et al.* 2009). In the field, relatively low numbers of tsetse flies (1–5%) are capable of successfully transmitting trypanosomes to a naïve mammalian host (Aksoy *et al.* 2003). Several physiological mechanisms have been reported that are involved in fly natural resistance to trypanosome infection. These include fly age and nutritional status (mostly teneral flies that are newly emerged and previously unfed are sensitive to infection) (Welburn and Maudlin 1992; Kubi *et al.* 2006; Walshe *et al.* 2011), antimicrobial peptides (Hao *et al.* 2001; Hu and Aksoy 2006), trypanosome-binding lectins (Welburn *et al.* 1994), gut-associated tsetse EP protein (Haines *et al.* 2010), reactive oxygen species (ROS) (MacLeod *et al.* 2007) and tsetse peptidoglycan recognition protein LB (PGRPLB) (Wang and Aksoy 2012).

If the balance between the ability of the fly to eliminate the infection and the capacity of the parasite to evade the hostile environment of the gut favors trypanosomes, after three to six days post infection, a proportion of procyclic parasites enters a differentiation program, characterized by progressive elongation, with further flagellar growth and increase in cell motility. These parasites are known as mesocyclic and cross the peritrophic matrix, colonize the ectoperitrophic space and reach the proventriculus (Figure 5.3) (Gibson and Bailey 2003). High densities of mesocyclic parasites are able to swim together, producing super-ordered and synchronized wave patterns and significant directional force (Schuster *et al.* 2017), which could fit their migratory function.

Glossina spp. tsetse flies produce a type II peritrophic matrix, a chitinous structure constitutively secreted from specialized enlarged epithelial cells in the proventriculus, which lines up with the fly midgut. It has been proposed that the matrix plays important roles in tsetse fly biology, such as the control of digestive enzyme movement into the lumen, preventing access to environmental toxins and mechanical damage caused from ingested food constituents; and the regulation of the tsetse's immune induction upon the parasite challenge (Hegedus *et al.* 2009; Weiss *et al.* 2013).

Importantly, forward motility impairment prevents trypanosomes from developing into proventricular stages, thus failing to complete their parasite cycle in the insect vector (Rotureau *et al.* 2014). Within the proventriculus (also called the "cardia"), long mesocyclic parasites become thinner and divide asymmetrically to produce a long and a short epimastigote cell (Figure 5.3) (Van den Abbeele *et al.* 1999; Sharma *et al.* 2008), an event that implies a drastic remodeling of the cytoskeleton and changes in the molecular composition and the organization of the flagellum attachment zone (Rotureau *et al.* 2011). While the kinetoplast localized midway between the cell nucleus and posterior in midgut forms, epimastigote parasites exhibit the kinetoplast anterior to the central nucleus.

Epimastigote parasites are thought to re-enter the midgut and migrate through the foregut to the salivary glands (Van den Abbeele *et al.* 1999). It is assumed that the long and highly motile epimastigote acts as a carrier that transports the short epimastigote cell into the lumen of the gland. The long cell presumably dies whereas the short one continues the cycle (Van den Abbeele *et al.* 1999; Sharma *et al.* 2008). This key step is one of the major bottlenecks for salivary gland infection and requires passage via the foregut and the hypopharynx and finally to the narrow salivary duct. To our knowledge, it has not been imaged directly, hence the pathway described above remains to be formally proven. There, the long and highly motile dividing epimastigote cell would be able to deliver the short epimastigote cell, whose motility properties appear more limited. It evidences the environmental adaptation to assemble and construct a long flagellum, as this elongated shape could offer less friction and facilitate migration (Rotureau *et al.* 2011). Here too, this reasonable hypothesis could not be formally challenged since mutants with reduced forward motility fail to reach the proventriculus (Rotureau *et al.* 2014), preventing further investigation.

In the salivary glands, the short epimastigote parasites attach to the epithelium, thanks to extensive outgrowths of the flagellar membrane between host microvilli to form close contacts and construct structures resembling hemidesmosomes (Tetley and Vickerman 1985). It should be noted that despite the morphological similarity, keratin and related proteins are missing from the trypanosome genome, indicating that unique components must have been evolved to mediate attachment. These cells express, on their surface, another stage-specific glycosylphosphatidyl inositol-anchored protein, known as BARPs (for *brucei* alanine-rich proteins; Urwyler *et al.* 2007), the only known molecular marker for this life stage to date. After attachment, these cells elongate their posterior end, adopting protrusions that oscillate in the lumen, offering a significant surface that could be involved in sensing and/or molecular exchanges with the environment (Figure 5.3). The repositioning of the kinetoplast in an anterior positioning relative to the nucleus facilitates extension of the free portion of the flagellum, allowing better substrate adherence.

Morphometric analyses performed in all stages of parasite development revealed that the evolution of cell length is closely related to variations in flagellum length (Rotureau *et al.* 2011; Ooi and Bastin 2013), thus indicating that modifications in flagellum length could drive morphological modifications as proposed during *in vitro* experiments manipulating flagellum length (Kohl *et al.* 2003). The evolution of cytoskeletal components *in vivo* in the tsetse fly appears to happen very differently between proliferative and differentiating stages, with a remodeling of the entire sub-pellicular cytoskeleton of the future short epimastigote that occurs before the asymmetric division takes place in the fly proventriculus (Rotureau *et al.* 2011). The cellular membrane of *T. brucei* is associated with this cytoskeleton scaffold by microtubule-associated proteins (MAPs) that contribute to the stability of the cytoskeleton (Hertz-Fowler *et al.* 2001; Olego-Fernandez *et al.* 2009; May *et al.* 2012). The distribution of these proteins is not homogeneous, since some MAPs are restricted to the anterior end (the less dynamic half of the corset), with differential expression and sub-cellular localization during the lifecycle (Vedrenne *et al.* 2002). Therefore, this tight and robust microtubule corset behaves as a flexible cellular structure providing a high degree of plasticity to allow for changes in cellular architecture.

A recent study revealed the existence of two types of proliferation that occur simultaneously in the salivary glands and that contribute to the continuous production of infective trypanosomes: the first cycle produces equivalent cells that are attached to the salivary gland epithelium and ensures the rapid colonization of the glands. The second type of proliferation relies on an asymmetric division that results in a daughter cell that matures into the infective metacyclic parasite that is released in the tsetse saliva

(Figure 5.3) (Rotureau *et al.* 2012). As metacyclic parasites recover the trypomastigote morphotype, it is fair to speculate that this recurrent morphological modification might ensure more efficient swimming in a multicellular and viscous environment, such as the bloodstream of the mammalian host (Schuster *et al.* 2017). During this last transition, metacyclic parasites remodel the protein repertoire on their surface and express variant surface glycoproteins (mVSG, from metacyclic VSG). Each metacyclic cell seems to activate randomly one from a small subset of *mVSG* genes, which results in a heterogeneous population, whereby each trypanosome expresses a single VSG (Dubois *et al.* 2005; Vigneron *et al.* 2020), and whose expression is controlled exclusively at the transcriptional level. The expression of different mVSGs by infective metacyclic cells is a protective pre-adaptation for the environment that the parasites encounter within the next developmental transition in the mammalian host. Importantly, once this metacyclic population has been established in the salivary glands, the tsetse fly can transmit parasites throughout its entire lifespan. Moreover, *T. brucei* infection in the salivary glands dramatically modulates the vector physiology and feeding behavior as a consequence of a changed protein content of the tsetse saliva (Van den Abbeele *et al.* 2010). The physical obstruction of the alimentary tract due to the presence of trypanosomes could reduce the arthropod feeding efficiency and therefore force tsetse flies to probe more frequently (Jenni *et al.* 1980; Van den Abbeele *et al.* 2010). Together with field studies that have proved that tsetse flies are particularly responsive to host defensive behavior and are prone to interrupted feeding (Schofield and Torr 2002), the increased probability of discontinued blood feeding would favor host/vector contact and, in turn, the chance of parasite transmission.

Progression of parasite infection in the fly also relies on parasite ability to adapt its metabolism to use the available nutrients. Besides the proteolytic and immune challenges, trypanosomes must adapt to the tsetse's midgut, which is rich in amino acids, particularly proline, which is the fly's hemolymph-borne energy source (Balogun 1974). Tsetse flies do not store carbohydrates for ATP production (Norden and Paterson 1969) and are adapted to metabolize amino acids. Thus, proline is used for ATP biosynthesis (Bursell 1960) and for important physiological activities, including lactation, fecundity and the flight process (Bursell 1973; Moloo 1976; Michalkova *et al.* 2014). In this context, glucose present within the blood meal will be rapidly metabolized (~1 h) and only minute amounts will be possibly used by the fly in order to generate specific metabolites. Although procyclic trypanosomes rely upon proline *in vivo* for ATP production, they prefer glucose in culture experiments, whereas this sugar is absent from the midgut lumen between blood meals (Lamour *et al.* 2005). Recently, RNA-seq analysis comparing procyclic and bloodstream parasites showed that the THT2 hexose transporters are transiently upregulated in early procyclic forms, which predominate during the establishment of an infection of the tsetse fly midgut, usually during the first four to seven days after ingestion (Naguleswaran *et al.* 2018). The authors speculated that it might reflect the need for active acquisition of glucose in a sugar-poor environment, providing a window for maturation of the mitochondrion. In addition, enzymes involved in amino acid catabolism are substantially less abundant in bloodstream forms, and it has been assumed that proline is not metabolized by these parasites (Lamour *et al.* 2005). Thus, once in the tsetse's midgut, bloodstream parasites must transit from a glucose-rich environment to a glucose-depleted one where amino acids become the major carbon source. ATP is then generated via oxidative phosphorylation within the parasite's mitochondrion upon differentiation (Lamour *et al.* 2005; Mantilla *et al.* 2017). Important morphological remodeling of the rudimentary mitochondrion present in bloodstream parasites is taking place in early procyclic cells, as this organelle is essential for ATP production through the citric acid cycle and the electron transport chain (Tielens and Van Hellemond 1998).

Furthermore, it has been demonstrated that in the absence of glucose, proline is essential for *T. brucei* survival in the tsetse vector, and that the proline metabolism pathway allows the parasite to adapt to the nutritional environment of the fly midgut (Mantilla *et al.* 2017). In addition to evident metabolic adaptations, mitochondrial enzymes also regulate the surface antigen expression. In early procyclic parasites, GPEET is the major surface protein, while in late procyclic forms, EP procyclin continues to be expressed (Vassella *et al.* 2000). Thus, GPEET procyclin shows alternative expression in association with the activity of mitochondrial enzymes, with downregulation of the acetate, succinate-CoA transferase/succinyl-CoA synthetase (ASCT) pathway (that converts acetyl-CoA to acetate and results in the formation of extra ATP; Van Hellemond *et al.* 1998), leading to expression of GPEET in late procyclic forms (Vassella *et al.* 2004).

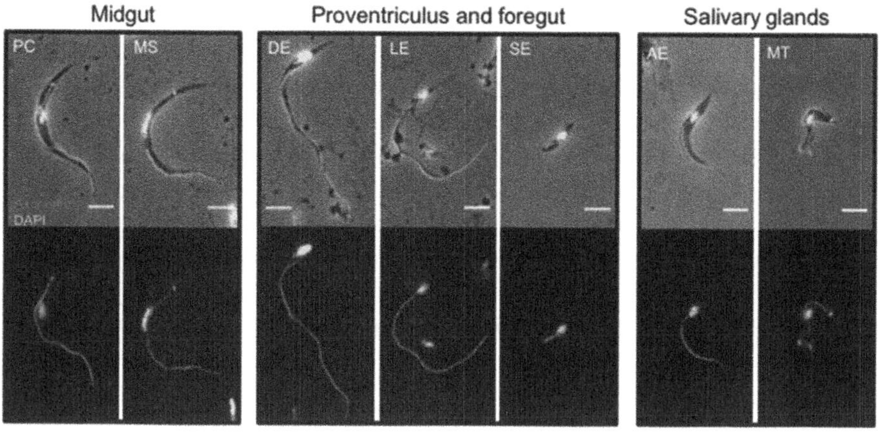

Figure 5.3. *Morphological changes in trypanosomes during the development in the insect vector. Representative pictures of each parasite stage in chronological order found in the midgut (left panel), the proventriculus and foregut (middle panel), and the salivary glands (right panel) are shown. Immunofluorescence images of methanol-fixed parasites stained with the anti-axoneme marker Mab25 (Pradel* et al. *2006) (magenta) and DAPI staining (cyan), labeling the flagellum and DNA, respectively. PC: procyclic trypomastigote; MS: mesocyclic trypomastigote; DE: asymmetrically dividing epimastigote; LE: long epimastigote; SE: short epimastigote; AE: attached epimastigote; MT: metacyclic trypomastigote. The scale bar is 5 μm (Estefanía Calvo Alvarez, unpublished results). For a color version of this figure, see www.iste.co.uk/grandcolas/ systematics.zip*

Importantly, the absence of glucose also implies that glucose-6-phosphate, a precursor for essential metabolic parasite pathways such as the pentose phosphate pathway and nucleotide sugar biosynthesis (Stokes *et al.* 2008; Kovářová and Barrett 2016), needs to be produced through another metabolic pathway, the so-called gluconeogenesis, and that key gluconeogenic enzymes are involved in parasite colonization of the salivary glands (Wargnies *et al.* 2018). This is the first example of a specific metabolic adaptation of *T. brucei* parasites needed for completion of the development in the insect vector.

We have seen that trypanosomes adapt the right differentiation program in the appropriate tissue. How could they detect the environment? The flagellum has been proposed to act as a sensory organelle (Rotureau *et al.* 2009a). This looks particularly relevant in the case of the attachment of epimastigote parasites to the salivary gland epithelium where extensive outgrowths of the flagellar membrane have been described (Tetley and Vickerman 1985). Indirect indications include the facts that flagellar membrane proteins can modulate virulence in the mammalian host (Webb *et al.* 1997; Emmer *et al.* 2010; Salmon *et al.* 2012), and flagellum-dependent signaling and sensing components, such as the mitogen-activated protein kinases (MAP) could be implicated in parasite environmental sensing, virulence and flagellar length control (Hua and Wang 1997; Muller *et al.* 2002; Domenicali-Pfister *et al.* 2006; Guttinger *et al.* 2007; Rotureau *et al.* 2009b). A recent study has proposed that flagellar cAMP signaling through the flagellar phosphodiesterase PDEB1 is required for traversal of the peritrophic matrix. Without PDEB1, parasites remain trapped in the midgut lumen (Shaw *et al.* 2019). These findings evidence the fact that throughout its cyclical development *T. brucei* might be able to sense its environment and to transduce and respond in order to reach the next compartment or to activate the next differentiation program.

In the future, highly sensitive metabolomic, transcriptomic or phosphoproteomic approaches could be crucial to get insights into the different metabolites present within the tissues of the insect vector, including the proventriculus and the salivary glands. Questions regarding the specific metabolic pathways and remodeling of epimastigote stages and infective metacyclic parasites are still unanswered. Although the crosstalk between the tsetse fly's and the trypanosome's metabolism is likely to be complex, the availability and turnover of energy sources will undoubtedly be the key component for parasite, and, thus, disease transmission.

5.4. Adaptations of *T. brucei* to the mammalian host

Another remarkable layer of complexity is present in the interaction between trypanosomes and the vertebrate host. The mammal becomes infected when a tsetse fly injects intradermally metacyclic parasites along with the anticoagulant saliva. Tsetse flies have a pool feeding strategy that allows the laceration of the skin with their proboscis (elongated, blood sucking part of the mouthparts) and blood ingestion from a superficial lesion (Lehane *et al.* 2004). Hereby, the skin of the vertebrate host is the first anatomical barrier that trypanosomes have to overcome in order to establish an infection. At the injection site, tsetse fly salivary components accelerate the onset of *T. brucei* infection through an immune-modulatory effect: pro-inflammatory cytokines are reduced at the site of infection initiation as well as serum concentrations of the trypanocidal host cytokine tumor necrosis factor (TNF-α) (Caljon *et al.* 2006).

Early at the dermal site of inoculation, metacyclic parasites differentiate into the so-called slender bloodstream stage, which are long (mean flagellar length of 27 μm), and proliferative parasites (Tyler *et al.* 2001) (Figure 5.4, left panel). The first adaptation that takes place is the remodeling of the metacyclic surface coat constituted of mVSG to form the bloodstream VSG cover, a dense coat that is composed of approximately 5×10^6 identical VSG homodimers that are attached to the plasma membrane by a GPI anchor, and that protects the parasites from host antibodies (Schwede *et al.* 2011, 2015).

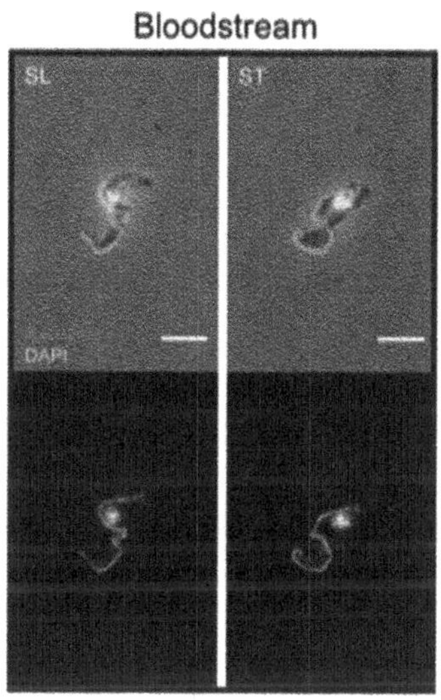

Figure 5.4. *Morphology of the main bloodstream form trypanosomes: proliferative slender (SL) and transmissible stumpy parasites (ST). Immunofluorescence of methanol-fixed parasites stained with the anti-axoneme antibody that recognizes the flagellum (magenta) and DAPI for DNA staining (cyan). The scale bar is 5 μm (Estefanía Calvo Alvarez, unpublished results). For a color version of this figure, see www.iste.co.uk/grandcolas/systematics.zip*

Later in the infection process, the presence of parasites in the brain reflects the invasion of the central nervous system, which causes the meningoencephalitis and neurological disorders typical of the second stage of the disease (Sternberg 2004). At the level of the host, massive splenomegaly (enlargement of the spleen), coincides with the destruction of the lymphoid tissue architecture and hepatomegaly (enlargement of the liver). Subsequent lymphadenopathy and hypergammaglobulinemia (overproduction of immunoglobulins in the blood) result in systemic multiple organ failure and death in experimental mouse models (Takeya *et al.* 1987).

T. brucei parasites have developed efficient evasion mechanisms, including modulating host defenses: they are master manipulators of the entire host immune system, leading to an ineffective host antibody response characterized by global immunosuppression (Magez *et al.* 1999). The extracellular lifestyle exhibited by *T. brucei* exposes the parasite to the immune system. Since infections are sustained for months to years, parasite persistence requires subsequent changes in cell morphology, metabolism, gene expression and signaling pathways.

In order to escape the adaptive immune system, proliferative slender trypanosomes, the ones which establish parasitemia, undergo antigenic variation during chronic infections (Cross 1990), a process that prevents parasite elimination by host antibodies raised against the immunodominant VSG. This mechanism relies on the large family of *VSG* genes (Berriman *et al.* 2005), a battery of approximately 2,000 genes and pseudogenes located in subtelomeric regions of the parasites' genome. In bloodstream forms, around 15 telomeric VSG expression sites contain eight to nine associated genes, 70 bp repeats and a *VSG* gene (Glover *et al.* 2013). *T. brucei* parasites exhibit monoallelic expression of one expression site at a time with the subsequent expression of a single VSG variant per parasite (Borst 2002). At regular time points during infection, some individual parasites switch their coat into a different antigen type, escaping from the humoral immune response against the prevalent (new) VSG species and ensuring the maintenance of a chronic infection (Figure 5.5). However, this key survival strategy is much greater than expected as "mosaic" variants (novel variants that form through gene conversion events within VSG sequences), containing at least 15 distinct antigen types, arise during individual growth peaks (Hall *et al.* 2013). Indeed, *VSG* genes can undergo segmental recombination to encode an estimated up to 10,000 different VSG surface coats during infection (Morrison *et al.* 2009). Nevertheless, little is known about the mechanisms by which these mosaic variants are formed and it has been suggested that "mosaic" VSG production might be partially inefficient, as it could produce non-functional VSGs (Mugnier *et al.* 2016). The continuous VSG coat turnover along with the sequential expression of unique and mosaic VSG coat proteins have long excluded the development of an effective vaccine against African trypanosomiasis.

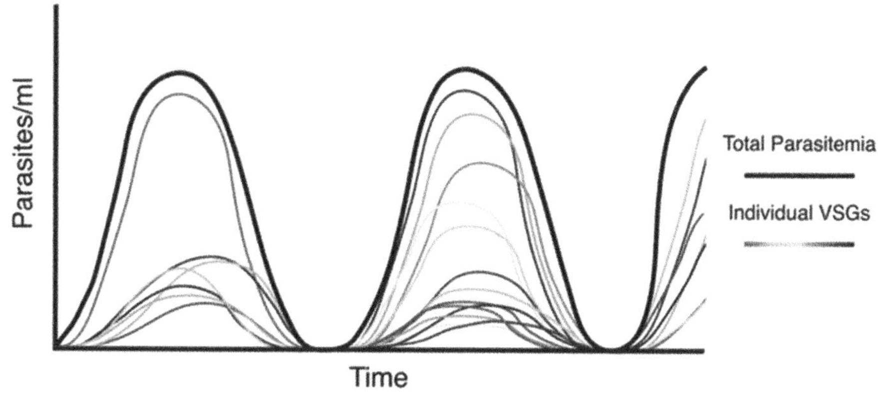

Figure 5.5. *VSG expression dynamics through* T. brucei *infection in mice. Each colored line represents an individual VSG variant expressed within a parasite population and the black line shows total parasitemia within the entire population (adapted from Mugnier* et al. *2016). For a color version of this figure, see www.iste.co.uk/grandcolas/systematics.zip*

In addition, the VSG coat functions to selectively remove antibodies bound to VSG protein from the cell surface. Rapid removal of host-derived immunoglobulin (Ig) stuck to endocytosed VSG occurs at the posterior cell pole, ensuring the elimination of the Ig-VSG immune complexes. This is carried out by hydrodynamic flow forces that result from forward cellular motility (Engstler *et al.* 2007).

The sustaining of chronic infections is achieved through the activation of another developmental program by slender parasites that ensures the survival of the mammalian host and potentially enhances the transmission capacity. As parasite numbers increase, slender parasites differentiate into cell cycle-arrested stumpy parasites, which appear shorter (mean flagellum length of 19 μm), and are non-proliferative stages (Figure 5.4, right panel) (MacGregor *et al.* 2012). This process depends on a quorum-sensing (QS) type response that occurs in a density-dependent manner resulting from the accumulation of a stumpy-induced factor produced by the parasite itself (Reuner *et al.* 1997; Vassella *et al.* 1997). Recently, the long-awaited mechanism of trypanosome quorum sensing has been elucidated: a surface transmembrane protein, functioning as a peptide transporter and expressed on slender parasites, regulates the differentiation process through oligopeptide signaling (Rojas *et al.* 2019).

The differentiation of slender parasites into the stumpy form is an adaptive mechanism that ensures the generation of pre-committed parasites that are capable of transmission and survival in the tsetse's midgut. This irreversible differentiation is accompanied by cell-cycle arrest in G1/G0 (Shapiro *et al.* 1984) that prolongs host survival and, hence, the probability of parasite transmission. Moreover, morphological changes that take place after transmission to the insect vector can be coordinated with re-entry into the cell cycle. This is of great importance as proper organelle positioning is essential for completion of the developmental cycle within the tsetse fly. An important feature is that stumpy parasites are able to detect the differentiation signal in the tsetse's midgut which comprises citrate/cis-aconitate (Czichos *et al.* 1986; Engstler and Boshart 2004). The expression of a carboxylate surface transporter family, called PAD proteins (Proteins Associated with Differentiation), allows stumpy parasites to detect this specific signal, in particular, PAD1, which is only expressed at significant levels in the transmission stage (Dean *et al.* 2009).

In bloodstream trypanosomes, important metabolic adaptations are essential to allow the parasites to re-program their metabolism in order to use the available nutrients within the different organs and tissues of the mammalian host. In contrast to tsetse-derived parasites, the single mitochondrion of slender trypanosomes is a simple tubular structure devoid of cristae. This reflects the absence of mitochondrial respiration and highlights the use of glucose as the only carbon source for ATP production. The presence of a stable glucose concentration within the bloodstream, as well as in the brain and other host tissues, shifts parasite metabolism towards the glycolytic pathway. Strikingly, the first seven glycolytic enzymes are compartmentalized in peroxisome-like organelles known as glycosomes (Opperdoes 1987; Gualdron-Lopez *et al.* 2012). To maintain the redox balance within the cell, parasites consume large amounts of oxygen in a reaction catalyzed by the trypanosome alternative oxidase, the unique enzyme used in mammalian stages for respiration. Another metabolic adaptation is the rate of glucose degradation, which is 10-fold higher in slender compared to procyclic parasites (Mazet *et al.* 2013). Compatible with this finding is the fact that some biological processes require more ATP, such as the endocytosis (Pal *et al.* 2002), which is important for the recycling of the VSG coat and for elimination of bound antibodies, as already discussed.

Although living in a glucose-rich environment, stumpy parasites undergo specific adaptations and modifications that will be needed to support the next differentiation steps within the tsetse fly's midgut environment (Kabani *et al.* 2009). Some of these include resistance to changes in temperature, as well as extracellular acidic and proteolytic stress (reviewed in Rico *et al.* 2013 and Nolan *et al.* 2000), components of the respiratory chain and other mitochondrial genes which are also upregulated in ST parasites (Feagin *et al.* 1986; Jensen *et al.* 2009; Kabani *et al.* 2009; Nilsson *et al.* 2010). These observations confirm that they are pre-adapted to the more complex metabolic requirements of life in the tsetse fly and highlight that these parasites are indeed prepared for transmission.

It was long though that parasites thrive mostly within the blood, hence the frequent description of "bloodstream" stages in the literature. However, recent data revealed that the skin and potentially the adipose tissues could be more relevant to the infection (Caljon *et al.* 2016; Capewell *et al.* 2016a, 2016b; Trindade *et al.* 2016). Remarkably, trypanosomes are present in the skin during the whole course of the infection in the mammalian host, while detectable levels of parasites in the blood seem to be more variable (Capewell *et al.* 2016a). It was already known that trypanosomes migrate to distinct host tissues and organs via the blood; however, this was the first report on the invasion and systemic spreading throughout the surface of the body.

The authors confirmed the striking relevance of the skin as an overlooked parasite reservoir. Thanks to bioluminescence imaging, they could detect areas of the mouse body where trypanosomes were present (Figure 5.6). When flies were allowed to probe in these areas, one-third of them were infected, even in circumstances where bloodstream parasites could not be detected. This experiment using natural vector transmission demonstrated that parasites in the skin can be transmitted to the vector, in agreement with the tsetse fly's peculiar mode of feeding. Moreover, examination of human skin biopsies of undiagnosed individuals revealed the presence of extravascular parasites (Capewell *et al.* 2016a). This discovery raises fundamental questions, as the presence of a significant transmissible population in the dermis of humans and animals could impact future control and elimination strategies for both HAT and animal trypanosomiases, especially as it points to the presence of parasites in apparently asymptomatic individuals (Informal Expert Group on Gambiense HAT Reservoirs *et al.* 2018).

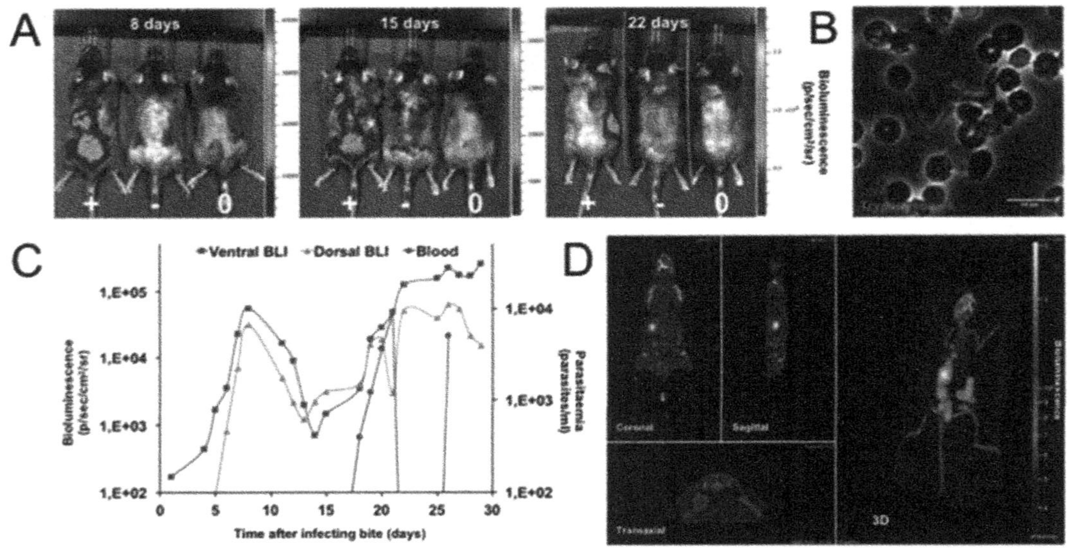

Figure 5.6. *Skin parasites in experimental* in vivo *infections.* T. brucei *parasites were engineered to express a chimeric triple reporter protein combining a "red-shifted" luciferase and a red fluorescent protein, fused by a small tag. A)* In vivo *whole-body images of C57BL/6J mice after a tsetse fly bite. The three animals represent one mouse bitten by an infected tsetse fly (+), one bitten by a non-infected fly (−) and one non-bitten (0). B) Immunofluorescence of methanol-fixed bloodstream parasites found in the blood of an infected mouse, labeled with an anti-mCherry antibody (red) and DNA staining (blue). C) Total bioluminescence signal emitted by the parasite-infected mouse, acquired within ventral (blue squares) and dorsal (green triangles) views, together with the parasitemia (red dots) for mouse +. D) Bioluminescence and anatomical 3D computed tomography images acquired by an IVIS Spectrum CT on a BALB/c mouse 29 days post-inoculation. The 3D reconstruction view on the right panel has been decomposed in pictures along three distinct plans showed on the left (coronal, sagittal and transaxial) (adapted from Calvo-Alvarez* et al. *2018). For a color version of this figure, see www.iste.co.uk/grandcolas/systematics.zip*

Finally, trypanosomes have also been detected in the adipose tissues in mouse models. Although the significance for transmission is unclear at this stage, it is noteworthy that parasites in the adipose tissue are transcriptionally distinct and upregulate genes for fatty acid metabolism, once more revealing metabolic adaptations to specific host microenvironments (Trindade *et al.* 2016).

5.5. Conclusion

During the course of evolution, *T. brucei* has been exposed to two hosts and showed remarkable abilities to adapt to very different environments in terms of temperature, viscous properties, energy sources and immune defenses. It dealt with these changing conditions by developing exquisite differentiation programs with appropriate and sometimes spectacular adaptations in metabolism, cell surface proteins or morphogenesis. But how are *T. brucei* parasites able to sense and detect the specific organ/tissue which they encounter in order to activate the next differentiation programs? The way the parasite detects these different environments remains enigmatic, although the contribution of the flagellum as a sensing organelle appears a promising lead for exciting future discoveries.

5.6. References

Adl, S.M., Simpson, A.G., Lane, C.E., Lukeš, J., Bass, D., Bowser, S.S., Brown, M.W., Burki, F., Dunthorn, M., Hampl, V., Heiss, A., Hoppenrath, M., Lara, E., Le Gall, L., Lynn, D.H., McManus, H., Mitchell, E.A., Mozley-Stanridge, S.E., Parfrey, L.W., Pawlowski, J., Rueckert, S., Shadwick, L., Schoch, C.L., Smirnov, A., and Spiegel, F.W. (2012). The revised classification of eukaryotes. *J. Eukaryot. Microbiol.*, 59(5), 429–493.

Aksoy, S., Gibson, W.C., and Lehane, M.J. (2003). Interactions between tsetse and trypanosomes with implications for the control of trypanosomiasis. *Adv. Parasitol.*, 53, 1–83.

Balogun, R.A. (1974). Studies on the amino acids of the tsetse fly, *Glossina morsitans*, maintained on *in vitro* and *in vivo* feeding systems. *Comp. Biochem. Physiol. A*, 49(2A), 215–222.

Benne, R., Van den Burg, J., Brakenhoff, J.P., Sloof, P., Van Boom, J.H., and Tromp M.C. (1986). Major transcript of the frameshifted coxII gene from trypanosome mitochondria contains four nucleotides that are not encoded in the DNA. *Cell*, 46(6), 819–826.

Benz, C., Mulindwa, J., Ouna, B., and Clayton, C. (2011). The *Trypanosoma brucei* zinc finger protein ZC3H18 is involved in differentiation. *Mol. Biochem. Parasitol.*, 177(2), 148–151.

Berriman, M., Ghedin, E., Hertz-Fowler, C. *et al.* (2005). The genome of the African trypanosome *Trypanosoma brucei*. *Science*, 309(5733), 416–422.

Borst, P. (2002). Antigenic variation and allelic exclusion. *Cell*, 109(1), 5–8.

Branche, C., Kohl, L., Toutirais, G., Buisson, J., Cosson, J., and Bastin, P. (2006). Conserved and specific functions of axoneme components in trypanosome motility. *J. Cell Sci.*, 119(16), 3443–3455.

Bursell, E. (1960). Free amino-acids of the tsetse fly (*Glossina*). *Nature*, 187, 778–727.

Bursell, E. (1973). Development of mitochondrial and contractile components of the flight muscle in adult tsetse flies, *Glossina morsitans*. *J. Insect Physiol.*, 19(5), 1079–1086.

Büscher, P., Cecchi, G., Jamonneau, V., and Priotto, G. (2017). Human African trypanosomiasis. *The Lancet*, 390(10110), 2397–2409.

Caljon, G., Van den Abbeele, J., Stijlemans, B., Coosemans, M., De Baetselier, P., and Magez, S. (2006). Tsetse fly saliva accelerates the onset of *Trypanosoma brucei* infection in a mouse model associated with a reduced host inflammatory response. *Infect. Immun.*, 74(11), 6324–6330.

Caljon, G., Van Reet, N., De Trez, C., Vermeersch, M., Perez-Morga, D., and Van den Abbeele, J. (2016). The dermis as a delivery site of *Trypanosoma brucei* for tsetse flies. *PLOS Pathog.*, 12(7), e1005744.

Calvo-Alvarez, E., Cren-Travaillé, C., Crouzols, A., and Rotureau, B. (2018). A new chimeric triple reporter fusion protein as a tool for *in vitro* and *in vivo* multimodal imaging to monitor the development of African trypanosomes and *Leishmania* parasites. *Infect. Genet. Evol.*, 63, 391–403.

Capewell, P., Cren-Travaillé, C., Marchesi, F., Johnston, P., Clucas, C., Benson, R.A., Gorman, T.A., Calvo-Alvarez, E., Crouzols, A., Jouvion, G., Jamonneau, V., Weir, W., Stevenson, M.L., O'Neill, K., Cooper, A., Swar, N.K., Bucheton, B., Ngoyi, D.M., Garside, P., Rotureau, B., and MacLeod, A. (2016). The skin is a significant but overlooked anatomical reservoir for vector-borne African trypanosomes. *eLife*, 5(September).

Cecchi, G., Mattioli, R.C., Slingenbergh, J., and De La Rocque, S. (2008). Land cover and tsetse fly distributions in sub-Saharan Africa. *Med. Vet. Entomol.*, 22(4), 364–373.

Clayton, C. (2013). The regulation of trypanosome gene expression by RNA-binding proteins. *PLOS Pathog.*, 9(11), e1003680.

Clayton, C. (2019). Regulation of gene expression in trypanosomatids: Living with polycistronic transcription. *Open Biol.*, 9(6), 190072.

Cooper, S., Wadsworth, E.S., Ochsenreiter, T., Ivens, A., Savill, N.J., and Schnaufer, A. (2019). Assembly and annotation of the mitochondrial minicircle genome of a differentiation-competent strain of *Trypanosoma brucei*. *Nucleic Acids Res.*, 47(21), 11304–11325.

Cross, G.A. (1990). Cellular and genetic aspects of antigenic variation in trypanosomes. *Annu. Rev. Immunol.*, 8, 83–110.

Czichos, J., Nonnengaesser, C., and Overath, P. (1986). *Trypanosoma brucei*: *cis*-aconitate and temperature reduction as triggers of synchronous transformation of bloodstream to procyclic trypomastigotes *in vitro*. *Exp. Parasitol.*, 62(2), 283–291.

Dacks, J.B., Walker, G., and Field, M.C. (2008). Implications of the new eukaryotic systematics for parasitologists. *Parasitol. Int.*, 57(2), 97–104.

Dean, S., Marchetti, R., Kirk, K., and Matthews, K.R. (2009). A surface transporter family conveys the trypanosome differentiation signal. *Nature*, 459(7244), 213–217.

Domenicali-Pfister, D., Burkard, G., Morand, S., Renggli, C.K., Roditi, I., and Vassella, E. (2006). A mitogen-activated protein kinase controls differentiation of bloodstream forms of *Trypanosoma brucei*. *Eukaryot. Cell*, 5(7), 1126–1135.

Dubois, M.E., Demick, K.P., and Mansfield, J.M. (2005). Trypanosomes expressing a mosaic variant surface glycoprotein coat escape early detection by the immune system. *Infect. Immun.*, 73(5), 2690–2697.

Emmer, B.T., Daniels, M.D., Taylor, J.M., Epting, C.L., and Engman, D.M. (2010). Calflagin inhibition prolongs host survival and suppresses parasitemia in *Trypanosoma brucei* infection. *Eukaryot. Cell*, 9(6), 934–942.

Engstler, M. and Boshart, M. (2004). Cold shock and regulation of surface protein trafficking convey sensitization to inducers of stage differentiation in *Trypanosoma brucei*. *Genes Dev.*, 18(22), 2798–2811.

Engstler, M., Pfohl, T., Herminghaus, S., Boshart, M., Wiegertjes, G., Heddergott, N., and Overath, P. (2007). Hydrodynamic flow-mediated protein sorting on the cell surface of trypanosomes. *Cell*, 131(3), 505–515.

Ersfeld, K. (2011). Nuclear architecture, genome and chromatin organisation in *Trypanosoma brucei*. *Res. Microbiol.*, 162(6), 626–636.

Feagin, J.E., Jasmer, D.P., and Stuart, K. (1986). Differential mitochondrial gene expression between slender and stumpy bloodforms of *Trypanosoma brucei*. *Mol. Biochem. Parasitol.*, 20(3), 207–214.

Field, M.C. and Carrington, M. (2009). The trypanosome flagellar pocket. *Nat. Rev. Microbiol.*, 7(11), 775–786.

Franco, J.R., Cecchi, G., Priotto, G., Paone, M., Diarra, A., Grout, L., Simarro, P.P., Zhao, W., and Argaw, D. (2020). Monitoring the elimination of human African trypanosomiasis: Update to 2018. *PLOS Negl. Trop. Dis.*, 14(5), e0008261.

Fulton, C. (1977). Cell differentiation in *Naegleria gruberi*. *Annu. Rev. Microbiol.*, 31, 597–629.

Gibson, W. and Bailey, M. (2003). The development of *Trypanosoma brucei* within the tsetse fly midgut observed using green fluorescent trypanosomes. *Kinetoplastid Biol. Dis.*, 2(1), 1.

Glover, L., Hutchinson, S., Alsford, S., Mcculloch, R., Field, M.C., and Horn, D. (2013). Antigenic variation in African trypanosomes: The importance of chromosomal and nuclear context in VSG expression control. *Cell. Microbiol.*, 15(12), 1984–1993.

Grandgenett, P.M., Otsu, K., Wilson, H.R., Wilson, M.E., and Donelson, J.E. (2007). A function for a specific zinc metalloprotease of African trypanosomes. *PLOS Pathog.*, 3(10), 1432–1445.

Gruszynski, A.E., van Deursen, F.J., Albareda, M.C., Best, A., Chaudhary, K., Cliffe, L.J., del Rio, L., Dunn, J.D., Ellis, L., Evans, K.J., Figueiredo, J.M., Malmquist, N.A., Omosun, Y., Palenchar, J.B., Prickett, S., Punkosdy, G.A., van Dooren, G., Wang, Q., Menon, A.K., Matthews, K.R., and Bangs, J.D. (2006). Regulation of surface coat exchange by differentiating African trypanosomes. *Mol. Biochem. Parasitol.*, 147(2), 211–223.

Gualdron-Lopez, M., Brennand, A., Hannaert, V., Quinones, W., Caceres, A.J., Bringaud, F., Concepcion, J.L., and Michels, P.A. (2012). When, how and why glycolysis became compartmentalised in the Kinetoplastea. A new look at an ancient organelle. *Int. J. Parasitol.*, 42(1), 1–20.

Guttinger, A., Schwab, C., Morand, S., Roditi, I., and Vassella, E. (2007). A mitogen-activated protein kinase of *Trypanosoma brucei* confers resistance to temperature stress. *Mol. Biochem. Parasitol.*, 153(2), 203–206.

Haines, L.R., Lehane, S.M., Pearson, T.W., and Lehane, M.J. (2010). Tsetse EP protein protects the fly midgut from trypanosome establishment. *PLOS Pathog.*, 6(3), e1000793.

Hall, J.P.J., Wang, H., and Barry, J.D. (2013). Mosaic VSGs and the scale of *Trypanosoma brucei* antigenic variation. *PLOS Pathog.*, 9(7), e1003502.

Hao, Z., Kasumba, I., Lehane, M.J., Gibson, W.C., Kwon, J., and Aksoy, S. (2001). Tsetse immune responses and trypanosome transmission: Implications for the development of tsetse-based strategies to reduce trypanosomiasis. *Proc. Natl. Acad. Sci. U.S.A*, 98(22), 12648–12653.

Hecker, H., Burri, H., Steiger, R., and Geigy, R. (1972). Morphometric data on the ultrastructure of the pleomorphic bloodforms of *Trypanosoma brucei*, Plimmer and Bradford, 1899. *Acta Trop.*, 29(2), 182–198.

Heddergott, N., Kruger, T., Babu, S.B., Wei, A., Stellamanns, E., Uppaluri, S., Pfohl, T., Stark, H., and Engstler, M. (2012). Trypanosome motion represents an adaptation to the crowded environment of the vertebrate bloodstream. *PLOS Pathog.*, 8(11), e1003023.

Hegedus, D., Erlandson, M., Gillott, C., and Toprak, U. (2009). New insights into peritrophic matrix synthesis, architecture, and function. *Annu. Rev. Entomol.*, 54, 285–302.

Hendriks, E.F. and Matthews, K.R. (2005). Disruption of the developmental programme of *Trypanosoma brucei* by genetic ablation of TbZFP1, a differentiation-enriched CCCH protein. *Mol. Microbiol.*, 57(3), 706–716.

Hendriks, E.F., Robinson, D.R., Hinkins, M., and Matthews, K.R. (2001). A novel CCCH protein which modulates differentiation of *Trypanosoma brucei* to its procyclic form. *EMBO J.*, 20(23), 6700–6711.

Hertz-Fowler, C., Ersfeld, K., and Gull, K. (2001). CAP5.5, a life-cycle-regulated, cytoskeleton-associated protein is a member of a novel family of calpain-related proteins in *Trypanosoma brucei. Mol. Biochem. Parasitol.*, 116(1), 25–34.

Hertz-Fowler, C., Figueiredo, L.M., Quail, M.A., Becker, M., Jackson, A., Bason, N., Brooks, K., Churcher, C., Fahkro, S., Goodhead, I., Heath, P., Kartvelishvili, M., Mungall, K., Harris, D., Hauser, H., Sanders, M., Saunders, D., Seeger, K., Sharp, S., Taylor, J.E., Walker, D., White, B., Young, R., Cross, G.A., Rudenko, G., Barry, J.D., Louis, E.J., and Berriman, M. (2008). Telomeric expression sites are highly conserved in *Trypanosoma brucei. PLOS One*, 3(10), e3527.

Hoare, C.A. and Wallace, F.G. (1966). Developmental stages of Trypanosomatid flagellates: A new terminology. *Nature*, 212(5068), 1385–1386.

Hong, M. and Simpson, L. (2003). Genomic organization of *Trypanosoma brucei* kinetoplast DNA minicircles. *Protist*, 154(2), 265–279.

Hu, C. and Aksoy, S. (2006). Innate immune responses regulate trypanosome parasite infection of the tsetse fly *Glossina morsitans morsitans*. *Mol. Microbiol.*, 60(5), 1194–1204.

Hua, S.B. and Wang, C.C. (1997). Interferon-gamma activation of a mitogen-activated protein kinase, KFR1, in the bloodstream form of *Trypanosoma brucei*. *J Biol. Chem.*, 272(16), 10797–10803.

Informal Expert Group on Gambiense HAT Reservoirs, Büscher, P., Bart, J.M., Boelaert, M. *et al.* (2018). Do cryptic reservoirs threaten gambiense-sleeping sickness elimination? *Trends Parasitol.*, 34(3), 197–207.

Jamonneau, V., Ilboudo, H., Kabore, J., Kaba, D., Koffi, M., Solano, P., Garcia, A., Courtin, D., Laveissiere, C., Lingue, K., Buscher, P., and Bucheton, B. (2012). Untreated human infections by *Trypanosoma brucei* gambiense are not 100% fatal. *PLOS Negl. Trop. Dis.*, 6(6), e1691.

Jenni, L., Molyneux, D.H., Livesey, J.L., and Galun, R. (1980). Feeding behaviour of tsetse flies infected with salivarian trypanosomes. *Nature*, 283(5745), 383–385.

Jensen, R.E. and Englund, P.T. (2012). Network news: The replication of kinetoplast DNA. *Annu. Rev. Microbiol.*, 66, 473–491.

Jensen, B.C., Sivam, D., Kifer, C.T., Myler, P.J., and Parsons, M. (2009). Widespread variation in transcript abundance within and across developmental stages of *Trypanosoma brucei*. *BMC Genom.*, 10, 482–482.

Kabani, S., Fenn, K., Ross, A., Ivens, A., Smith, T.K., Ghazal, P., and Matthews, K. (2009). Genome-wide expression profiling of *in vivo*-derived bloodstream parasite stages and dynamic analysis of mRNA alterations during synchronous differentiation in *Trypanosoma brucei*. *BMC Genom.*, 10, 427–427.

Kohl, L., Robinson, D., and Bastin, P. (2003). Novel roles for the flagellum in cell morphogenesis and cytokinesis of trypanosomes. *EMBO J.*, 22(20), 5336–5346.

Kolev, N.G., Ramey-Butler, K., Cross, G.A., Ullu, E., and Tschudi, C. (2012). Developmental progression to infectivity in *Trypanosoma brucei* triggered by an RNA-binding protein. *Science*, 338(6112), 1352–1353.

Kovářová, J. and Barrett, M.P. (2016). The pentose phosphate pathway in parasitic trypanosomatids. *Trends Parasitol.*, 32(8), 622–634.

Krenn, H.W. and Aspöck, H. (2012). Form, function and evolution of the mouthparts of blood-feeding Arthropoda. *Arthropod Struct. Dev.*, 41(2), 101–118.

Kubi, C., Van den Abbeele, J., De Dekken, R., Marcotty, T., Dorny, P., and Van den Bossche, P. (2006). The effect of starvation on the susceptibility of teneral and non-teneral tsetse flies to trypanosome infection. *Med. Vet. Entomol.*, 20(4), 388–392.

Lamour, N., Rivière, L., Coustou, V., Coombs, G.H., Barrett, M.P., and Bringaud, F. (2005). Proline metabolism in procyclic *Trypanosoma brucei* is down-regulated in the presence of glucose. *J. Biol. Chem.*, 280(12), 11902–11910.

Lehane, M.J., Aksoy, S., and Levashina, E. (2004). Immune responses and parasite transmission in blood-feeding insects. *Trends Parasitol.*, 20(9), 433–439.

Lukeš, J., Wheeler, R., Jirsová, D., David, V., and Archibald, J.M. (2018). Massive mitochondrial DNA content in diplonemid and kinetoplastid protists. *IUBMB Life*, 70(12), 1267–1274.

MacGregor, P. and Matthews, K.R. (2010). New discoveries in the transmission biology of sleeping sickness parasites: Applying the basics. *J. Mol. Med. (Berlin, Germany)*, 88(9), 865–871.

MacGregor, P., Szoor, B., Savill, N.J., and Matthews, K.R. (2012). Trypanosomal immune evasion, chronicity and transmission: An elegant balancing act. *Nat. Rev. Microbiol.*, 10(6), 431–438.

MacLeod, E.T., Maudlin, I., Darby, A.C., and Welburn, S.C. (2007). Antioxidants promote establishment of trypanosome infections in tsetse. *Parasitology*, 134(6), 827–831.

Magez, S., Radwanska, M., Beschin, A., Sekikawa, K., and De Baetselier, P. (1999). Tumor necrosis factor alpha is a key mediator in the regulation of experimental *Trypanosoma brucei* infections. *Infect. Immun.*, 67(6), 3128–3132.

Mantilla, B.S., Marchese, L., Casas-Sánchez, A., Dyer, N.A., Ejeh, N., Biran, M., Bringaud, F., Lehane, M.J., Acosta-Serrano, A., and Silber, A.M. (2017). Proline metabolism is essential for *Trypanosoma brucei brucei* survival in the tsetse vector. *PLOS Pathog.*, 13(1), e1006158.

Matthews, K.R. (1999). Developments in the differentiation of *Trypanosoma brucei. Parasitol. Today*, 15(2), 76–80.

May, S.F., Peacock, L., Almeida Costa, C.I., Gibson, W.C., Tetley, L., Robinson, D.R., and Hammarton, T.C. (2012). The *Trypanosoma brucei* AIR9-like protein is cytoskeleton-associated and is required for nucleus positioning and accurate cleavage furrow placement. *Mol. Microbiol.*, 84(1), 77–92.

Mazet, M., Morand, P., Biran, M., Bouyssou, G., Courtois, P., Daulouède, S., Millerioux, Y., Franconi, J.-M., Vincendeau, P., Moreau, P., and Bringaud, F. (2013). Revisiting the central metabolism of the bloodstream forms of *Trypanosoma brucei*: Production of acetate in the mitochondrion is essential for parasite viability. *PLoS Negl. Trop. Dis.*, 7(12), e2587.

Michaeli, S. (2011). Trans-splicing in trypanosomes: Machinery and its impact on the parasite transcriptome. *Future Microbiol.*, 6(4), 459–474.

Michalkova, V., Benoit, J.B., Weiss, B.L., Attardo, G.M., and Aksoy, S. (2014). Vitamin B6 generated by obligate symbionts is critical for maintaining proline homeostasis and fecundity in tsetse flies. *Appl. Environ. Microbiol.*, 80(18), 5844–5853.

Moloo, S.K. (1976). Aspects of the nutrition of adult female *Glossina morsitans* during pregnancy. *J. Insect Physiol.*, 22(4), 563–567.

Morrison, L.J., Majiwa, P., Read, A.F., and Barry, J.D. (2005). Probabilistic order in antigenic variation of *Trypanosoma brucei*. *Int. J. Parasitol.*, 35(9), 961–972.

Morrison, L.J., Marcello, L., and McCulloch, R. (2009). Antigenic variation in the African trypanosome: Molecular mechanisms and phenotypic complexity. *Cell Microbiol.*, 11(12), 1724–1734.

Mowatt, M.R. and Clayton, C.E. (1987). Developmental regulation of a novel repetitive protein of *Trypanosoma brucei*. *Mol. Cell. Biol.*, 7(8), 2838–2844.

Mugnier, M.R., Stebbins, C.E., and Papavasiliou, F.N. (2016). Masters of disguise: Antigenic variation and the VSG coat in *Trypanosoma brucei*. *PLOS Pathog.*, 12(9), e1005784.

Mugo, E. and Clayton, C. (2017). Expression of the RNA-binding protein RBP10 promotes the bloodstream-form differentiation state in *Trypanosoma brucei*. *PLOS Pathog.*, 13(8), e1006560.

Muller, I.B., Domenicali-Pfister, D., Roditi, I., and Vassella, E. (2002). Stage-specific requirement of a mitogen-activated protein kinase by *Trypanosoma brucei*. *Mol. Biol. Cell*, 13(11), 3787–3799.

Naguleswaran, A., Doiron, N., and Roditi, I. (2018). RNA-Seq analysis validates the use of culture-derived *Trypanosoma brucei* and provides new markers for mammalian and insect life-cycle stages. *BMC Genom.*, 19(1), 227–227.

Nilsson, D., Gunasekera, K., Mani, J., Osteras, M., Farinelli, L., Baerlocher, L., Roditi, I., and Ochsenreiter, T. (2010). Spliced leader trapping reveals widespread alternative splicing patterns in the highly dynamic transcriptome of *Trypanosoma brucei*. *PLOS Pathog.*, 6(8), e1001037.

Nolan, D.P., Rolin, S., Rodriguez, J.R., Van den Abbeele, J., and Pays, E. (2000). Slender and stumpy bloodstream forms of *Trypanosoma brucei* display a differential response to extracellular acidic and proteolytic stress. *Eur. J. Biochem.*, 267(1), 18–27.

Norden, D.A. and Paterson, D.J. (1969). Carbohydrate metabolism in flight muscle of the tsetse fly (*Glossina*) and the blowfly (*Sarcophaga*). *Comp. Biochem. Physiol.*, 31(5), 819–827.

Ogbadoyi, E.O., Robinson, D.R., and Gull, K. (2003). A high-order trans-membrane structural linkage is responsible for mitochondrial genome positioning and segregation by flagellar basal bodies in trypanosomes. *Mol. Biol. Cell*, 14(5), 1769–1779.

Olego-Fernandez, S., Vaughan, S., Shaw, M.K., Gull, K., and Ginger, M.L. (2009). Cell morphogenesis of *Trypanosoma brucei* requires the paralogous, differentially expressed calpain-related proteins CAP5.5 and CAP5.5V. *Protist*, 160(4), 576–590.

Ooi, C.P. and Bastin, P. (2013). More than meets the eye: Understanding *Trypanosoma brucei* morphology in the tsetse. *Front. Cell. Infect. Microbiol.*, 3, 71.

Opperdoes, F.R. (1987). Compartmentation of carbohydrate metabolism in trypanosomes. *Annu. Rev. Microbiol.*, 41, 127–151.

Pal, A., Hall, B.S., Nesbeth, D.N., Field, H.I., and Field, M.C. (2002). Differential endocytic functions of *Trypanosoma brucei* Rab5 isoforms reveal a glycosyl-phosphatidylinositol-specific endosomal pathway. *J. Biol. Chem.*, 277(11), 9529–9539.

Pradel, L.C., Bonhivers, M., Landrein, N., and Robinson, D.R. (2006). NIMA-related kinase TbNRKC is involved in basal body separation in *Trypanosoma brucei*. *J. Cell Sci.*, 119(9), 1852–1863.

Ralston, K.S., Kabututu, Z.P., Melehani, J.H., Oberholzer, M., and Hill, K.L. (2009). The *Trypanosoma brucei* flagellum: Moving parasites in new directions. *Annu. Rev. Microbiol.*, 63, 335–362.

Reuner, B., Vassella, E., Yutzy, B., and Boshart, M. (1997). Cell density triggers slender to stumpy differentiation of *Trypanosoma brucei* bloodstream forms in culture. *Mol. Biochem. Parasitol.*, 90(1), 269–280.

Rico, E., Rojas, F., Mony, B.M., Szoor, B., MacGregor, P., and Matthews, K.R. (2013). Bloodstream form pre-adaptation to the tsetse fly in *Trypanosoma brucei*. *Front. Cell. Infect. Microbiol.*, 3, 78.

Robinson, D.R. and Gull, K. (1991). Basal body movements as a mechanism for mitochondrial genome segregation in the trypanosome cell cycle. *Nature*, 352(6337), 731–733.

Robinson, D.R., Sherwin, T., Ploubidou, A., Byard, E.H., and Gull, K. (1995). Microtubule polarity and dynamics in the control of organelle positioning, segregation, and cytokinesis in the trypanosome cell cycle. *J. Cell Biol.*, 128(6), 1163–1172.

Robinson, N.P., Burman, N., Melville, S.E., and Barry, J.D. (1999). Predominance of duplicative VSG gene conversion in antigenic variation in African trypanosomes. *Mol. Cell. Biol.*, 19(9), 5839–5846.

Roditi, I., Carrington, M., and Turner, M. (1987). Expression of a polypeptide containing a dipeptide repeat is confined to the insect stage of *Trypanosoma brucei*. *Nature*, 325(6101), 272–274.

Rojas, F., Silvester, E., Young, J., Milne, R., Tettey, M., Houston, D.R., Walkinshaw, M.D., Pérez-Pi, I., Auer, M., Denton, H., Smith, T.K., Thompson, J., and Matthews, K.R. (2019). Oligopeptide signaling through TbGPR89 drives trypanosome quorum sensing. *Cell*, 176(1–2), 306–317.

Rotureau, B., Morales, M.A., Bastin, P., and Späth, G.F. (2009). The flagellum-mitogen-activated protein kinase connection in Trypanosomatids: A key sensory role in parasite signalling and development? *Cell. Microbiol.*, 11(5), 710–718.

Rotureau, B., Subota, I., and Bastin, P. (2011). Molecular bases of cytoskeleton plasticity during the *Trypanosoma brucei* parasite cycle. *Cell Microbiol.*, 13(5), 705–716.

Rotureau, B., Subota, I., Buisson, J., and Bastin, P. (2012). A new asymmetric division contributes to the continuous production of infective trypanosomes in the tsetse fly. *Development*, 139(10), 1842–1850.

Rotureau, B., Ooi, C.P., Huet, D., Perrot, S., and Bastin, P. (2014). Forward motility is essential for trypanosome infection in the tsetse fly. *Cell Microbiol.*, 16(3), 425–433.

Salmon, D., Vanwalleghem, G., Morias, Y., Denoeud, J., Krumbholz, C., Lhommé, F., Bachmaier, S., Kador, M., Gossmann, J., Dias, F.B., De Muylder, G., Uzureau, P., Magez, S., Moser, M., De Baetselier, P., Van den Abbeele, J., Beschin, A., Boshart, M., and Pays, E. (2012). Adenylate cyclases of *Trypanosoma brucei* inhibit the innate immune response of the host. *Science*, 337(6093), 463–466.

Savage, A.F., Cerqueira, G.C., Regmi, S., Wu, Y., El Sayed, N.M., and Aksoy, S. (2012). Transcript expression analysis of putative *Trypanosoma brucei* GPI-anchored surface proteins during development in the tsetse and mammalian hosts. *PLoS Negl. Trop. Dis.*, 6(6), e1708.

Schofield, S. and Torr, S.J. (2002). A comparison of the feeding behaviour of tsetse and stable flies. *Med. Vet. Entomol.*, 16(2), 177–185.

Schuster, S., Krüger, T., Subota, I., Thusek, S., Rotureau, B., Beilhack, A., and Engstler, M. (2017). Developmental adaptations of trypanosome motility to the tsetse fly host environments unravel a multifaceted *in vivo* microswimmer system. *eLife*, 6, e27656.

Schwede, A., Jones, N., Engstler, M., and Carrington, M. (2011). The VSG C-terminal domain is inaccessible to antibodies on live trypanosomes. *Mol. Biochem. Parasitol.*, 175(2), 201–204.

Schwede, A., MacLeod, O.J., MacGregor, P., and Carrington, M. (2015). How does the VSG coat of bloodstream form African trypanosomes interact with external proteins? *PLOS Pathog.*, 11(12), e1005259.

Shapiro, S.Z., Naessens, J., Liesegang, B., Moloo, S.K., and Magondu, J. (1984). Analysis by flow cytometry of DNA synthesis during the life cycle of African trypanosomes. *Acta Trop.*, 41(4), 313–323.

Sharma, R., Peacock, L., Gluenz, E., Gull, K., Gibson, W., and Carrington, M. (2008). Asymmetric cell division as a route to reduction in cell length and change in cell morphology in trypanosomes. *Protist*, 159(1), 137–151.

Sharma, R., Gluenz, E., Peacock, L., Gibson, W., Gull, K., and Carrington, M. (2009). The heart of darkness: Growth and form of *Trypanosoma brucei* in the tsetse fly. *Trends Parasitol.*, 25(11), 517–524.

Shaw, S., Demarco, S.F., Rehmann, R., Wenzler, T., Florini, F., Roditi, I., and Hill, K.L. (2019). Flagellar cAMP signaling controls trypanosome progression through host tissues. *Nat. Commun.*, 10(1), 803–803.

Sherwin, T. and Gull, K. (1989). The cell division cycle of *Trypanosoma brucei brucei*: Timing of event markers and cytoskeletal modulations. *Philos. Trans. R. Soc. Lond. B Biol. Sci.*, 323(1218), 573–588.

Sherwin, T., Schneider, A., Sasse, R., Seebeck, T., and Gull, K. (1987). Distinct localization and cell cycle dependence of COOH terminally tyrosinolated alpha-tubulin in the microtubules of *Trypanosoma brucei brucei*. *J. Cell Biol.*, 104(3), 439–446.

Simpson, L. (1987). The mitochondrial genome of kinetoplastid protozoa: Genomic organization, transcription, replication, and evolution. *Annu. Rev. Microbiol.*, 41, 363–382.

Simpson, A.G.B. and Roger, A.J. (2004). The real "kingdoms" of eukaryotes. *Curr. Biol.*, 14(17), R693–R696.

Singla, V. and Reiter, J.F. (2006). The primary cilium as the cell's antenna: Signaling at a sensory organelle. *Science*, 313(5787), 629–633.

Sternberg, J.M. (2004). Human African trypanosomiasis: Clinical presentation and immune response. *Parasite Immunol.*, 26(11–12), 469–476.

Stokes, M.J., Güther, M.L.S., Turnock, D.C., Prescott, A.R., Martin, K.L., Alphey, M.S., and Ferguson, M.A.J. (2008). The synthesis of UDP-N-acetylglucosamine is essential for bloodstream form *Trypanosoma brucei in vitro* and *in vivo* and UDP-N-acetylglucosamine starvation reveals a hierarchy in parasite protein glycosylation. *J. Biol. Chem.*, 283(23), 16147–16161.

Subota, I., Rotureau, B., Blisnick, T., Ngwabyt, S., Durand-Dubief, M., Engstler, M., and Bastin, P. (2011). ALBA proteins are stage regulated during trypanosome development in the tsetse fly and participate in differentiation. *Mol. Biol. Cell*, 22(22), 4205–4219.

Takeya, M., Reinwald, E., and Risse, H.J. (1987). Pathobiochemical alterations in experimental chronic and acute trypanosomal infection in mice. *J. Clin. Chem. Clin. Biochem.*, 25(10), 665–673.

Tetley, L. and Vickerman, K. (1985). Differentiation in *Trypanosoma brucei*: Host–parasite cell junctions and their persistence during acquisition of the variable antigen coat. *J. Cell Sci.*, 74, 1–19.

Tielens, A.G. and Van Hellemond, J.J. (1998). Differences in energy metabolism between trypanosomatidae. *Parasitol. Today*, 14(7), 265–272.

Trindade, S., Rijo-Ferreira, F., Carvalho, T., Pinto-Neves, D., Guegan, F., Aresta-Branco, F., Bento, F., Young, S.A., Pinto, A., Van den Abbeele, J., Ribeiro, R.M., Dias, S., Smith, T.K., and Figueiredo, L.M. (2016). *Trypanosoma brucei* parasites occupy and functionally adapt to the adipose tissue in mice. *Cell Host Microbe*, 19(6), 837–848.

Tyler, K.M., Matthews, K.R., and Gull, K. (2001). Anisomorphic cell division by African trypanosomes. *Protist*, 152(4), 367–378.

Urwyler, S., Studer, E., Renggli, C.K., and Roditi, I. (2007). A family of stage-specific alanine-rich proteins on the surface of epimastigote forms of *Trypanosoma brucei*. *Mol. Microbiol.*, 63(1), 218–228.

Van den Abbeele, J., Claes, Y., Van Bockstaele, D., Le Ray, D., and Coosemans, M. (1999). *Trypanosoma brucei* spp. development in the tsetse fly: Characterization of the post-mesocyclic stages in the foregut and proboscis. *Parasitology*, 118(5), 469–478.

Van den Abbeele, J., Caljon, G., De Ridder, K., De Baetselier, P., and Coosemans, M. (2010). *Trypanosoma brucei* modifies the tsetse salivary composition, altering the fly feeding behavior that favors parasite transmission. *PLOS Pathog.*, 6(6), e1000926.

Van Hellemond, J.J., Opperdoes, F.R., and Tielens, A.G. (1998). Trypanosomatidae produce acetate via a mitochondrial acetate: Succinate CoA transferase. *Proc. Natl. Acad. Sci. U.S.A*, 95(6), 3036–3041.

Vassella, E., Reuner, B., Yutzy, B., and Boshart, M. (1997). Differentiation of African trypanosomes is controlled by a density sensing mechanism which signals cell cycle arrest via the cAMP pathway. *J. Cell Sci.*, 110(21), 2661–2671.

Vassella, E., Van den Abbeele, J., Butikofer, P., Renggli, C.K., Furger, A., Brun, R., and Roditi, I. (2000). A major surface glycoprotein of Trypanosoma brucei is expressed transiently during development and can be regulated post-transcriptionally by glycerol or hypoxia. Genes Dev., 14(5), 615–626.

Vassella, E., Probst, M., Schneider, A., Studer, E., Renggli, C.K., and Roditi, I. (2004). Expression of a major surface protein of *Trypanosoma brucei* insect forms is controlled by the activity of mitochondrial enzymes. *Mol. Biol. Cell*, 15(9), 3986–3993.

Vaughan, S. (2010). Assembly of the flagellum and its role in cell morphogenesis in *Trypanosoma brucei*. *Curr. Opin. Microbiol.*, 13(4), 453–458.

Vaughan, S., Kohl, L., Ngai, I., Wheeler, R.J., and Gull, K. (2008). A repetitive protein essential for the flagellum attachment zone filament structure and function in *Trypanosoma brucei*. *Protist*, 159(1), 127–136.

Vedrenne, C., Giroud, C., Robinson, D.R., Besteiro, S., Bosc, C., Bringaud, F., and Baltz, T. (2002). Two related subpellicular cytoskeleton-associated proteins in *Trypanosoma brucei* stabilize microtubules. *Mol. Biol Cell*, 13(3), 1058–1070.

Vickerman, K. (1965). Polymorphism and mitochondrial activity in sleeping sickness trypanosomes. *Nature*, 208(5012), 762–766.

Vigneron, A., O'Neill, M.B., Weiss, B.L., Savage, A.F., Campbell, O.C., Kamhawi, S., Valenzuela, J.G., and Aksoy, S. (2020). Single-cell RNA sequencing of *Trypanosoma brucei* from tsetse salivary glands unveils metacyclogenesis and identifies potential transmission blocking antigens. *Proc. Natl. Acad. Sci. U.S.A*, 117(5), 2613–2621.

Walker, P.J. (1961). Organization of function in trypanosome flagella. *Nature*, 189, 1017–1018.

Walrad, P.B., Capewell, P., Fenn, K., and Matthews, K.R. (2012). The post-transcriptional trans-acting regulator, TbZFP3, co-ordinates transmission-stage enriched mRNAs in *Trypanosoma brucei*. *Nucleic Acids Res.*, 40(7), 2869–2883.

Walshe, D.P., Lehane, M.J., and Haines, L.R. (2011). Post eclosion age predicts the prevalence of midgut trypanosome infections in *Glossina*. *PLOS One*, 6(11), e26984.

Wang, J. and Aksoy, S. (2012). PGRP-LB is a maternally transmitted immune milk protein that influences symbiosis and parasitism in tsetse's offspring. *Proc. Natl. Acad. Sci. U.S.A*, 109(26), 10552–10557.

Wargnies, M., Bertiaux, E., Cahoreau, E., Ziebart, N., Crouzols, A., Morand, P., Biran, M., Allmann, S., Hubert, J., Villafraz, O., Millerioux, Y., Plazolles, N., Asencio, C., Rivière, L., Rotureau, B., Boshart, M., Portais, J.-C., and Bringaud, F. (2018). Gluconeogenesis is essential for trypanosome development in the tsetse fly vector. *PLOS Pathog.*, 14(12), e1007502.

Webb, H., Carnall, N., Vanhamme, L., Rolin, S., Van den Abbeele, J., Welburn, S., Pays, E., and Carrington, M. (1997). The GPI-phospholipase C of *Trypanosoma brucei* is nonessential but influences parasitemia in mice. *J. Cell Biol.*, 139(1), 103–114.

Webster, P. and Russell, D.G. (1993). The flagellar pocket of trypanosomatids. *Parasitol. Today*, 9(6), 201–206.

Weiss, B.L., Wang, J., Maltz, M.A., Wu, Y., and Aksoy, S. (2013). Trypanosome infection establishment in the tsetse fly gut is influenced by microbiome-regulated host immune barriers. *PLOS Pathog.*, 9(4), e1003318.

Welburn, S.C. and Maudlin, I. (1992). The nature of the teneral state in *Glossina* and its role in the acquisition of trypanosome infection in tsetse. *Ann. Trop. Med. Parasitol.*, 86(5), 529–536.

Welburn, S.C., Maudlin, I., and Molyneux, D.H. (1994). Midgut lectin activity and sugar specificity in teneral and fed tsetse. *Med. Vet. Entomol.*, 8(1), 81–87.

Welburn, S.C., Molyneux, D.H., and Maudlin, I. (2016). Beyond tsetse – Implications for research and control of human African trypanosomiasis epidemics. *Trends Parasitol.*, 32(3), 230–241.

Zhou, Q., Liu, B., Sun, Y., and He, C.Y. (2011). A coiled-coil- and C2-domain-containing protein is required for FAZ assembly and cell morphology in *Trypanosoma brucei*. *J. Cell Sci.*, 124(22), 3848–3858.

6

Challenges Inherent in the Systematics and Taxonomy of Genera that have Recently Experienced Explosive Radiation: The Case of Orchids of the Genus *Ophrys*

Orchids of the genus *Ophrys* have experienced recent explosive evolutionary radiation. Over the last few million years, these orchids have developed a sexual deception strategy allowing them to adapt to various pollinating insects (especially wild bees) that have existed for much longer. Within this genus, a large number of species that have been described by orchidophiles have been slow to be validated by the scientific community. Most of these taxa are interfertile, which makes the application of the "biological" concept of the species delicate.

Although the ten or so species constituting the "flagships" of the genus have been confirmed by various phylogenetic analyses, the delineation of the remaining 350 or so species, described to date, remains rather muddled. In this context, where the taxonomic activity of orchidophiles tends to be faster than that of scientists, recent advances in genomics and integrative taxonomy open up promising perspectives. Scientific validation of the limits between species within the genus *Ophrys* could allow for a better understanding of the diversity of this emblematic genus by the general public, as well as allow

---

Chapter written by Joris BERTRAND, Michel BAGUETTE, Nina JOFFARD and Bertrand SCHATZ.

the implementation of conservation actions that are more effective because they are better targeted.

6.1. Introduction

Within the Orchidaceae family, the genus *Ophrys* presents a monophyly that is strongly supported by a bundle of genetic, phenotypic and ecological data. All species of the genus *Ophrys* share a common strategy called "sexual swindle" to attract their pollinating insect(s). This pollination strategy is a key biological innovation to explain their "explosive" radiation. Their flowers look strangely similar to the female of the wild bee species that pollinates them. This resemblance is not only morphological (size, shape and sometimes hairiness) but, above all, olfactory since the odor emitted by the labellum (differentiated central petal on which the pollinator is positioned) is very similar to the sex pheromone specific to the bee species (see the review of Baguette *et al.* (2020)). Generally emerging one to two weeks before the females, the males of the bee species in question are thus confronted with these particular flowers, in the absence of females. Attracted to these flowers, they try to copulate with the labellum and, as they move, stick the pollinia to their bodies. By repeating this behavior on other flowers of the same species of *Ophrys*, they spread the pollen from flower to flower. *Ophrys* thus parasitize the specific attraction of the male to the female bee, which explains why each species of *Ophrys* is predominantly pollinated by a single bee species (Joffard *et al.* (2019) and references therein). Wild bees diversified in the Cretaceous 110–140 million years ago, in parallel with the diversification of Angiosperms (Danforth *et al.* 2013). For its part, the genus *Ophrys* only appeared about five million years ago (Breitkopf *et al.* 2015) and then experienced episodes of massive diversification through pollinator transitions, allowing both the exploitation of new ecological niches and the establishment of *reproductive isolation* between lineages. It is this historical context, and the asymmetrical relationship between *Ophrys* and pollinators that, to a large extent, explain the spectacular diversification of this genus. Considering that almost 2,000 species of wild bees are recorded in Europe (Nieto *et al.* 2014) – not counting those from North Africa – the Euro-Mediterranean genus *Ophrys* could diversify even further on the basis of quasi-specific pollination interactions.

Since the genus *Ophrys* has undergone significant, rapid and recent diversification, the systematics and, *a fortiori*, the taxonomy of this genus is far from having reached a state of consensus. Thus, while Delforge (2016) lists no less than 354 species of *Ophrys* in the most recent version of his book *Orchids of Europe, North Africa and the Near East*, some authors acknowledge the relevance of only nine species and are relatively skeptical about the quantity of what they consider to be "ethological" species (Bateman 2018; Bateman *et al.* 2018). Most scientists working on the genus *Ophrys* will claim to be more interested in understanding the ecological and evolutionary processes that have led to the impressive diversity of this genus than in delineating species within it. In fact, it is the very definition of the species that leads these scientists to position themselves in this range, varying from a dozen to several hundred species of *Ophrys*. They are therefore responsible for arguing this position, especially since it can have important implications, notably in terms of conservation biology. The relative democratization of high-throughput sequencing and metabolomic approaches, as well as the emergence of integrative taxonomy methods, now allows the combined analysis of genetic and phenotypic information (morphology and floral odors, for example). These approaches seem particularly promising in terms of being able to delineate *Ophrys* species in a reproducible and updatable way, based on statistical procedures as decision support tools.

6.2. Speciation in *Ophrys*: an evolutionary divergence seen as a reticulated continuum

6.2.1. *Difficulty in applying the biological concept of the species in the case of* Ophrys

The species has traditionally been defined as a group of natural populations that are effectively or potentially interferent and are genetically isolated from other similar groups. This biological concept of the species, formulated by Mayr (1942), appears to be relatively unsuitable for plants in that even occasional formation of hybrids (which may be viable and fertile) would concern at least a quarter of the species recognized as valid (Mallet 2005). This situation is all the more characterized within the genus *Ophrys*, in which it appears that the vast majority of species are interfertile if *prezygotic isolation* is circumvented (see, for example, Scopece *et al.* (2007)).

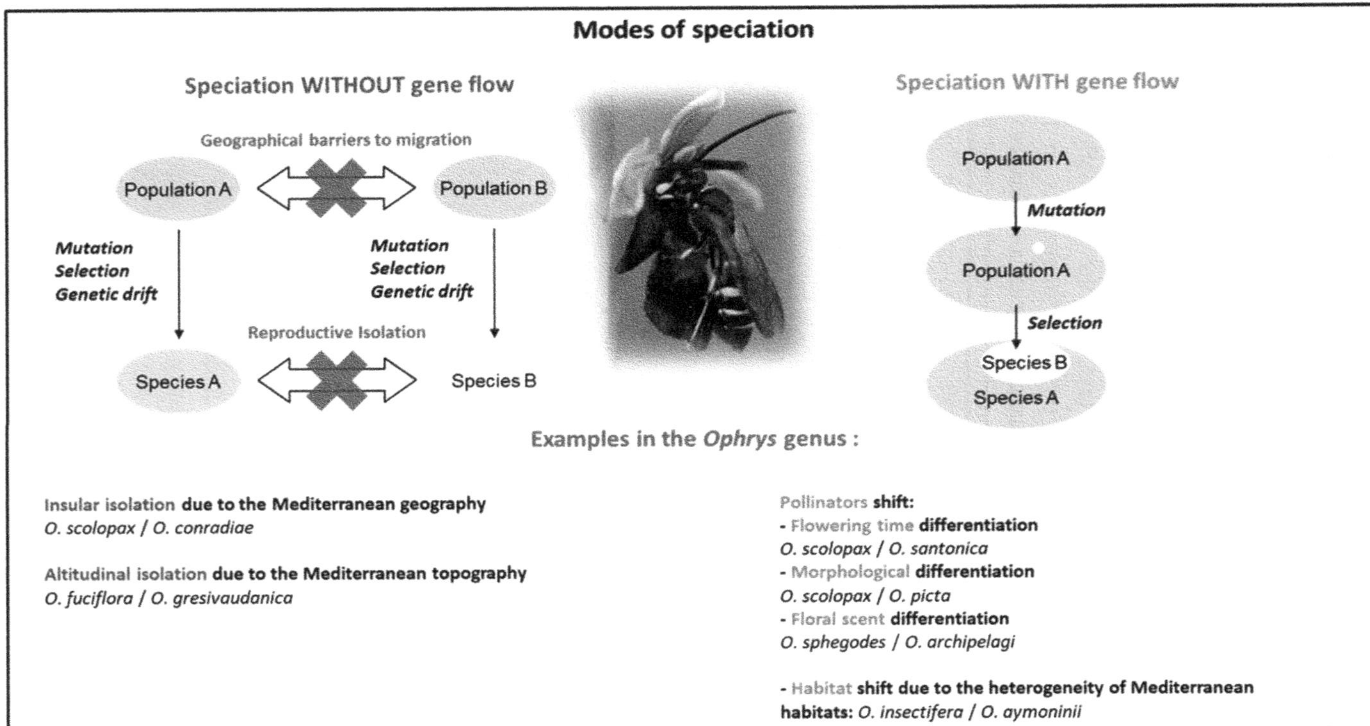

Figure 6.1. *Summary of the different speciation processes known in the genus* Ophrys *(photo by S. Witzthum). For a color version of this figure, see www.iste.co.uk/grandcolas/systematics.zip*

In order to propose a coherent definition of the notion of species for the *Ophrys*, it is therefore essential to consider the species in the context of the process of speciation. Speciation is generally defined as the formation of new species that diverge from common ancestors. This evolutionary divergence involves reproductive isolation corresponding to the emergence of *pre-* or *postzygotic* barriers that theoretically result in a definitive cessation of gene flow between lineages.

6.2.2. *Causes of reproductive isolation in* Ophrys

Reproductive isolation (Figure 6.1) is all the easier to understand as it materializes by an immediate, total and permanent cessation of gene flow between lineages, simultaneously initiating the process of speciation. Populations can thus find themselves completely genetically isolated from each other by a physical barrier that cannot be crossed by the propagules responsible for the flow of genes (seeds or pollen): this is the phenomenon of *allopatric speciation*, vicariant or geographical. This mode of speciation – often regarded as the default mode – is characterized by an absence of migration, so that no gene flow counteracts the effect of other evolutionary forces, such as mutation, natural selection and genetic drift on either side of the physical barrier. This mode of speciation is said to be relatively widespread in *Ophrys*, where no less than 45 species have been recorded as endemic to Mediterranean island systems, whether within large islands or within smaller islands such as Malta, Karpathos, Andros or Rhodes – bearing in mind that this number does not include species present on several groups of islands (Schatz 2017). For example, the group of *Ophrys bertolonii* is represented in the Balearic Islands by *O. balearica*, the group of *O. lunulata* in Malta by *O. melitensis* and that of *O. bornmuelleri* in Andros by *O. andria*. These species of *Ophrys* are endemic to these islands, on which they probably differed as a result of their geographical isolation. It should be noted that their main pollinating insect is not endemic to these islands. This differentiation can, if necessary, be amplified by the combined effects of adaptation to the particular constraints of an island ecology and genetic drift, all the more expected in the case where the island lineage would have formed from small numbers. In continental environments, cases of vicariance are generally more difficult to document, but the observation of *Ophrys* taxa with disjointed distribution areas (*O. santonica*, among others, present in Saintonge (Poitou-Charentes) and southern France) highlights and confirms the importance of geographical factors in speciation.

On the contrary, evolutionary divergence can also occur in the absence of geographical isolation and be of a purely ecological nature. The very notion of *sympatric speciation* has long been debated. However, in the particular case of *Ophrys*, it is now accepted that reproductive isolation could quite frequently be initiated in response to ecological isolation alone (reviewed by Baguette *et al.* (2020)). A growing number of studies thus supports a scenario in which mutations modifying the odor bouquet emitted by an individual *Ophrys*, in a locally rare or new way, could give it an adaptive advantage in its ability to attract a new pollinator species. This crossing of peaks in the olfactory landscape of the guild of pollinating insects, similar to the peaks of the selective value (*fitness*) in the adaptive landscape proposed by Wright (1932), would make it possible to initiate the implementation of prezygotic reproductive isolation. The directional selection imposed by the new pollinator could subsequently strengthen the reproductive barrier by shaping the floral morphology of the emerging lineage and/or by inducing phenological shifts (in other words, the flowering period), which also result in temporal reproductive isolation. Thus, some groups, such as that of *O. scolopax*, present very similar taxa that have different pollinators and differ by flower size (which are, for example, smaller in *O. picta*) or by non-overlapping flowering periods (which are, for example, later in *O. santonica*). It should be noted that although habitat preferences are less important, they may also induce reproductive isolation of an ecological (or geographical) nature in *Ophrys*. While most *Ophrys* species thrive in calcareous soils, some can tolerate a slightly acidic pH substrate (such as *O. scolopax*, *O. conradiae* and *O. tenthredinifera*, among those in metropolitan France). To ensure germination, all *Ophrys* must also establish a mycorrhizal symbiosis with habitat-dependent soil fungi in the seed state. However, to our knowledge, there are no studies demonstrating that these mycorrhizae are involved in reproductive isolation between *Ophrys* species, a topic that has not been fully explored (Roche *et al.* 2010; Schatz *et al.* 2010; Gervasi *et al.* 2017).

6.2.3. *Consequences of the implementation of reproductive isolation in the particular case of the genus* Ophrys

The difficulty in placing the concept of species in the two previous scenarios is that the discontinuation of gene flow may not be a total cessation but rather a gradual occurrence, only beginning to be detectable on the traits (and genes) that are involved in *adaptation* and reproductive isolation (Figure 6.2). This consideration is important because it allows us to adopt a less categorical view of speciation, which can then be seen as a continuum ranging from evolutionary divergence in the total absence of gene flows to more nuanced scenarios, in which reproductive isolation is initially partial and then, eventually, consolidates over time. In addition, reproductive isolation can be dynamic, both in time and space. In *Ophrys*, as in other plants and animals, some species have partially overlapping (*parapatric*) areas of distribution. This is the case, for example, for widely distributed species such as *O. insectifera*, which are locally in contact with endemic taxa with a much more restricted distribution of the same group, such as *O. aymoninii* or *O. subinsectifera*, that are occasionally forming small, but potentially stable *hybrid zones* in space and time (Triponez *et al.* 2013; Gervasi *et al.* 2017). In other taxonomic groups, it has been shown that such a situation would often result from *secondary contact* between two taxa that had previously diverged allopatrically, rather than from incipient parapatric divergence. Hybrid zones are thus generally the scene of the resumption of gene flow – be it sometimes partial and temporary – following the geographic expansion of the area of distribution of at least one of the two species.

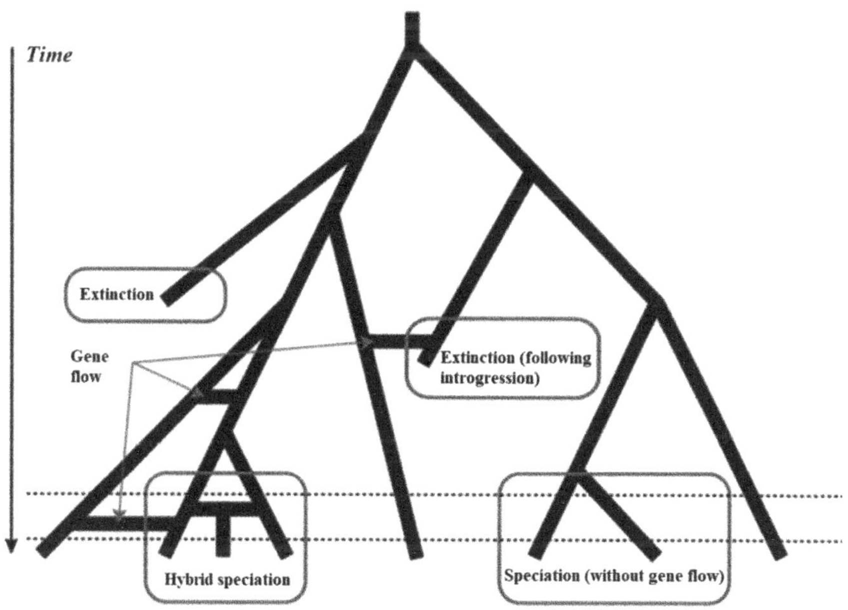

Figure 6.2. *Speciation in a context of reticulated evolution*

COMMENT ON FIGURE 6.2. – *The right part illustrates a "traditional" evolutionary divergence without gene flow during speciation. The left part illustrates the possible consequences of reticulated evolution. Introgressive hybridization can sometimes be at the origin of new evolutionary lineages (hybrid speciation), can sometimes threaten the integrity of pre-existing lineages and sometimes take place without perceptibly altering the integrity of the lineages. The positioning of the "taxonomic cursor" (dotted line) determines the taxonomic rank of the lineages: here, it is 6 or 8 according to the criteria.*

All of these cases of divergence involving a reduced, but not zero, gene flow (synonymous with incomplete reproductive isolation) thus attest to a potentially *reticulated* evolution, in which lineages would start to diverge from each other before potentially merging again, leading to the emergence of a new hybrid lineage (*hybrid speciation*) or a *hybrid swarm*, in addition to at least one of the two parental lineages. It should be noted that hybridization does not necessarily generate diversity, such as when the integrity of at least one of the parental lineages is threatened by the support of hybrids and *introgression* with the parental lineages. When species produce hybrids that are non-viable, sterile, or generally less able to survive and reproduce, the existence of postzygotic barriers is then assumed. In *Ophrys*, current knowledge indicates that the majority of hybrids are viable (Scopece *et al.* 2007) but may be sterile (Vereecken *et al.* 2010) or, at any rate, have selective value (*fitness*) that is often less than that of the parental taxa (Cortis *et al.* 2009).

The notion of species is therefore *a priori* subjective in the sense that it depends on a "minimal" divergence threshold, on which the "taxonomic cursor" is positioned on the continuum of evolutionary divergence. The positioning of this cursor is, of course, all the more difficult in a context of reticulated evolution. In the case of the study of speciation in *Ophrys*, it seems relevant to consider the evolutionary lineage as a unit of work, from the moment it shows (multiple) signals of pre- or postzygotic isolation. Thus, two very close taxa, showing simultaneous – even subtle – differences at the morphological and/or phenological level(s), and which are obviously attracting different pollinator species, could be considered to be at the early stages of the speciation process. In this context, the unifying definition of the species, based on evolutionary dynamics in metapopulations (de Queiroz 2005, 2007), is particularly interesting as it presents the double advantage of conceptual relevance and operational efficiency in terms of conservation. As the divergence between lineages is recent and a reticulated evolution is very likely, it seems inappropriate to strictly adopt the concept of phylogenetic species in the genus *Ophrys*, as proposed by some authors (Bateman 2018). This concept aims to recognize mutually monophyletic groups (clades) that are significantly divergent from each other (Devillers and Devillers-Terschuren 2013) as valid sister species. In the case of *Ophrys*, most of the proposed phylogenetic hypotheses have, so far, only been based on a handful of genes at most, each containing too few signals to clearly discriminate between species. The realization of studies at the phylogenomic scale should make it possible to obtain a more consensual phylogeny of the genus, and will at least have the merit of better understanding the degree of reticulation of its evolution. In the meantime, we suggest relying on an accumulation of evidence of reproductive isolation in an integrative approach, in order to define taxa with the main objective of identifying and describing speciation-relevant lineages, in other words, those engaged in a process of evolutionary divergence.

6.3. Current state of knowledge on *Ophrys* systematics

6.3.1. *Molecular systematics: overview of current knowledge*

Studies based on molecular phylogenies have greatly contributed to clarify (1) the positioning of *Ophrys* among the other genera of the Orchidaceae family and (2) the relationships between lineages within the genus. The genus *Ophrys* was first considered as a sister genus of a group including the genera *Anacamptis* (*sensu lato*) and *Serapias* (Pridgeon *et al.* 1997; Bateman *et al.* 2003) and then of a group including *Anacamptis* (*sensu lato*) and *Serapias*, as well as *Himantoglossum* (Inda *et al.* 2012). Even more recently, a study also including the species *Steveniella satyrioides* suggests that this monospecific genus could be the sister genus of *Ophrys*, these two genera forming the sister group of another clade including *Anacamptis* (*sensu lato*), *Serapias* and *Himantoglossum* (Jin *et al.* 2017). While the relationships between the genera *Ophrys*, *Anacamptis*, *Himantoglossum*, *Serapias* – and to a lesser extent *Steveniella* – have now been corroborated by work using significantly different genetic markers and methods, the connection of this clade to other genera of the *Orchidineae* subtribe remains uncertain.

The common denominator of these different studies is the use of *Internal Transcribed Spacers* 1 and 2 (ITS1/ITS2) and the gene for the small 5.8 S subunit of ribonucleic acid, a region of ribosomal DNA located between the small and large subunits (18 S and 26 S, respectively). This portion of DNA, carried by certain chromosomes (on the nuclear genome), is repeated in tandem a large number of times. This large amount of template DNA facilitates its amplification and sequencing. In addition, the subunits of ribosomal RNA are essential to the proper functioning of any organism, so that the strong stabilizing selection pressures exerted on the genes that code for them ensure the relative homogeneity of these sequences between species that are, from a phylogenetic point of view, relatively distant. They thus provide favorable conserved regions to define "universal" primers, in order to target an amplification of ITS regions that are variable and informative from a phylogenetic point of view. Indeed, ITS sequences are not subject to such selective constraints and can thus accumulate more mutations without significantly impacting the individuals carrying them. Such a level of polymorphism provides more information to infer phylogenetic relationships between species that have only diverged quite recently. Other advantages of this portion of DNA for the study of the molecular phylogeny of *Ophrys* have also been explained by Bateman (2018).

It should be noted that ITS sequences may, on the contrary, prove to be unusable for reconstructing phylogeny at higher taxonomic levels (e.g. at the family level) because they evolve too rapidly to infer the homology of molecular traits. For example, ITS sequences of species belonging to the genera *Ophrys* and *Cypripedium*, whose common ancestor would have lived more than 75 million years ago, are, in practice, impossible to align unambiguously. To obtain more information, preferably from *loci* independent of ITSs, other authors such as Inda *et al.* (2012) have included plastid (*rpl16*) and mitochondrial (*cox1*) *loci*, while Jin *et al.* (2017) included both a larger number of plastid *loci* (*matK*, *psaB*, *rbcl*, *trnL-F*, *trnH-psba*) and an additional nuclear gene, that of xanthine dehydrogenase (*Xdh*). Since these studies focused on higher taxonomic levels, none of them proposed an exhaustive taxonomic sampling of the genus *Ophrys*.

The molecular phylogeny of the genus *Ophrys* was also explored in more detail. As early as 2001, Soliva *et al.* proposed a phylogeny including 32 species and based on ITSs and the intronic chloroplast sequence *trnL-F*. Although the relationships between the different species remain uncertain, this study highlights the existence of three main lineages that do not quite correspond to the *Euophrys/Pseudophrys* dichotomy that has traditionally been used to classify *Ophrys*, on the basis of morpho-ecological criteria.

The terms *Euophrys* and *Pseudophrys* classify *Ophrys* according to the part of the insect's body on which the pollinia are deposited during pseudocopulation. By definition, *Pseudophrys* correspond to *Ophrys* in which the pollinia are deposited on the abdominal region of the pollinating insect, while Euophrys correspond to species in which the pollinia are deposited on the cephalic region of the pollinator. The work of Soliva *et al.* (2001), like the majority of subsequent studies, establishes that *Pseudophrys* would belong to a common lineage with the *O. bombyliflora*, *O. tenthredinifera* and *O. speculum* groups; this lineage being itself a sister of a lineage including the *O. insectifera* group, on the one side, and another lineage including all the other *Euophrys* ("recent" *Euophrys* (Tyteca and Baguette 2017)), on the other side. In contrast, the "archaic" *Euophrys* form a polyphyletic group including both the *O. insectifera* group and the *O. bombyliflora*, *O. speculum* and *O. tenthredinifera* groups.

This topology will be partly contradicted by later studies which sometimes position the *O. insectifera* group as a sister group to the *Pseudophrys* and other "archaic" *Euophrys* (Bateman *et al.* 2003, 2018), sometimes in a basal position (Breitkopf *et al.* 2015; Devey *et al.* 2018), where other studies have failed to make a decision on the position of this group (Tyteca and Baguette 2017). The study by Devey *et al.* (2008) extends the taxonomic sampling by incorporating 131 species whose sequencing of the ITSs and chloroplastic regions *trnD-trnT* and *trnH-psbA* will notably clarify the relationships between macro-species (as defined by Bateman *et al.* (2018)) or *Ophrys* "group head" species (10 groups (noted from A to J), which are easily distinguishable on the basis of the markers used, such as floral morphology). It should be noted that the work of Tyteca and Baguette (2017) will have the merit of incorporating 107 species but on an even larger number of individuals (189), even though this work is based on the sequencing of ITSs alone.

To date, the most widely used phylogenetic hypothesis remains that proposed by Breitkopf *et al.* (2015). It includes 37 species of *Ophrys*, but is innovative in that it has incorporated five nuclear markers, developed on the basis of the first published genomic resource for the genus, namely, the sequence of a transcriptome of *O. sphegodes* (Sedeek *et al.* 2013). Like the amplification of ITSs, the principle of the development of these genetic markers consists of targeting the amplification of supposedly variable intronic regions, a few hundred nucleotides long, by anchoring primers on adjacent exonic regions, which are supposedly conserved between close species. These *loci* were designated by the authors as: *beta-galactosidase-like* (BGP), *cinnamoyl alcohol dehydrogenase* (CAD), *fatty acid desaturase* (FAD), *putative long-chain acyl-CoA synthase* (LACS) and *transcription factor-like* (MYB). In addition, the authors incorporated LEAFY/FLORICULA (LFY), a single-copy nuclear *locus* that has previously been shown to be useful in the genus *Ophrys* (Schlüter *et al.* 2007). This study also supports the hypothesis that the genus *Ophrys* has three main clades: an "A clade" (including *O. insectifera*), a "B clade" (including the groups *O. speculum*, *O. bombyliflora*, *O. tenhtredinifera* and *O. fusca*) and a "C clade" including all the other *Euophrys* (*O. apifera*, *O. heldreichii*, *O. umbilicata*, *O. scolopax*, *O. holoserica* (syn. *O. fuciflora*) and *O. sphegodes* (the last three being difficult to distinguish)).

6.3.2. *Molecular systematics in the age of phylogenomics*

The most recent phylogenetic hypothesis concerning the genus *Ophrys* has been published by Bateman *et al.* (2018). It "only" includes 32 species of *Ophrys*, but can be considered as the first *phylogenomic* study for the genus, as it is based on a matrix of 4,159 genetic variants (single nucleotide polymorphisms, SNPs) which constitute as many traits obtained by a genome subsampling approach of RADseq (restriction site-associated DNA sequencing). Since the *Ophrys* genome is too large to be fully simultaneously sequenced on a satisfactory number of individuals, the principle of this approach first consists of digesting the total genomic DNA by restriction enzymes. These endonucleases cut the DNA as soon as they recognize a particular pattern on its sequence. For example, the restriction enzyme EcoRI comes into action as soon as it is in the presence of the six GAATTC nucleotides, but other enzymes will recognize other patterns, sometimes of different sizes, that are specific to them. The ends of these fragments are then sequenced using high-throughput sequencing techniques and the reads are processed to provide genetic markers that are comparable between individuals, populations and species. Although the name RADseq is often used as a generic term for a multitude of protocols, all of them involve at least one step of genome digestion by at least one restriction enzyme, as well as sequencing of the resulting fragments using high-throughput sequencing methods (sometimes also referred to as "genotyping by sequencing methods"). The results of the study by Bateman *et al.* (2018) are generally consistent with the previous results of Breitkopf *et al.* (2015), with the notable exception of the position of the *O. insectifera* clade.

Furthermore, the study by Bateman *et al.* clarifies the systematic relationships between the different clades of *Ophrys* (called "macro-species" by Bateman *et al.* (2018)).

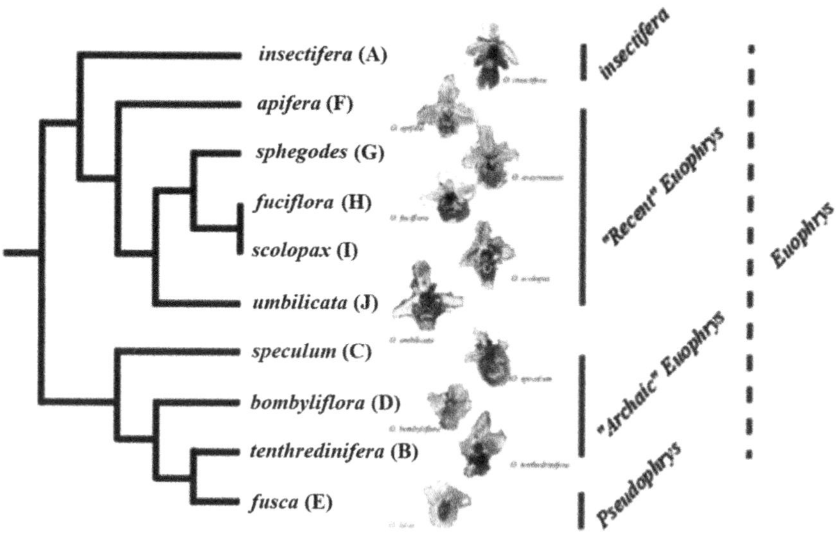

Figure 6.3. *Most likely topology to represent the phylogenetic relationships between Ophrys "flagship" species (macro-species) in light of the most recent genomic data (photos by J. Bertrand, B. Schatz and "Orchi" (Wikimedia Commons)). For a color version of this figure, see www.iste.co.uk/grandcolas/systematics.zip*

6.4. Integrative genomics and taxonomy: perspectives and issues

6.4.1. *Moving towards a generalization of data sets at the genomic scale*

Like many other non-model organisms, the *Ophrys* system is currently benefiting from the relative democratization of high-throughput sequencing methods. Pioneering work such as that of Sedeek *et al.* (2014), on an "intraspecific" scale, or Bateman *et al.* (2018), on an "interspecific" scale, should thus become more widespread in the coming years. The study of systems undergoing such "explosive" evolutionary diversification phenomena requires the use of a large number of genetic markers to (1) hopefully have sufficient resolution to infer informative and reliable phylogenetic hypotheses and (2) to compare the different topologies obtained from these thousands of genetic markers, in order to understand the degree of reticulation of the evolutionary tree of *Ophrys*, particularly by trying to distinguish events of gene flow (introgression) from those of *incomplete lineage sorting*. To do this, some opt for the use of tools derived from population genetics *a priori* dedicated to the study of genetic diversity and differentiation on an intraspecific scale: this is, for example, the case of RAD-type sequencing approaches (or genotyping by sequencing).

If these approaches prove to be powerful in solving the evolutionary history of close lineages, they will reach different limits each time at higher taxonomic scales. One of the main drawbacks of this type of approach is that DNA modifications explain that the restriction sites at which enzymes cut DNA are more likely to vary the more evolutionarily divergent individuals are genotyped. As a result, the amount of missing data in the matrix tends to increase as increasingly divergent lineages are incorporated, thus decreasing the size of the dataset. An alternative would be to use global methods which essentially consist of randomly sequencing portions of the genome, with the aim of assembling as many new fragments as possible in the hope of finding homologous regions between individuals.

These approaches have recently made it possible to successfully sequence and assemble relatively repeated portions of the *Ophrys* genome at a controlled cost, such as those in the chloroplasts of *O. sphegodes* and *O. fusca iricolor* (Roma *et al.* 2018) or *O. aveyronensis* (Bertrand *et al.* 2019). The sophistication of these approaches also makes it possible to target specific areas of the genome; however, their development requires extensive prior knowledge of the genome. To date, there are only four transcriptomes (Piñeiro Fernández *et al.* 2019) and no published reference genome for *Ophrys*. This relative delay in the generation of genomic resources available for *Ophrys* can largely be explained by the large size of the genomes of these plants (Leitch *et al.* 2009). Moreover, more chloroplast genomes and transcriptomes are expected to be published in the near future, as well as the first reference genome. As soon as this gap is filled, all the elements will be available to implement sequence capture approaches that can be used to simultaneously sequence hundreds or even thousands of genomic regions.

6.4.2. *Integrative taxonomy approach*

Taxonomic confusions within the genus *Ophrys* have significant consequences in terms of conservation. Several of its species are classified as "data deficient" in the IUCN Red Lists, including in France (IUCN *et al.* 2010), in particular due to uncertainties regarding their taxonomic rank. For this reason, there is an urgent need to redefine the boundaries between *Ophrys* species, in order to improve knowledge of their conservation status and the appropriation of these boundaries by stakeholders in the field.

As mentioned above, not all scientists (1) have the same definition of what a species is, (2) use the same criteria to distinguish between species (Bateman *et al.* 2011; Vereecken *et al.* 2011) and (3) require the same level of differentiation between two taxa in order to consider them as species (in other words, the opposition between lumpers and splitters (Pedersen and Faurholdt 2007; Delforge 2016)). In particular, there is a debate (Bateman *et al.* 2011; Vereecken *et al.* 2011) among proponents of "morphogenetic" criteria, who argue that taxa should have achieved reciprocal monophyly and/or be characterized by diagnostic morphological traits to be considered "good" species (Devey *et al.* 2008), and proponents of "eco-ethological" criteria, who argue that interactions between *Ophrys* and pollinators are more informative than neutral markers because of their key role in speciation (Schiestl and Ayasse 2002; Ayasse *et al.* 2011).

At the same time, orchidophiles continue to describe new species of *Ophrys* on essentially morphological grounds, as well as elevate varieties or subspecies to the rank of species, a phenomenon known as "taxonomic inflation" (Isaac *et al.* 2004). This strong activity by orchidophiles results in the publication of works aimed at describing new species, often without any real scientific justification. At the same time, scientists have been slow to re-evaluate the relevance of each of these new *Ophrys* species because of the complex logistics of sampling to be carried out throughout the Mediterranean basin and the costs of analysis. This explains the perpetual debate surrounding the identification of *Ophrys* and a constructive outcome must be found to this debate.

In this context, the application of a so-called "integrative" taxonomy approach, which would make use of both morphogenetic and eco-ethological criteria, seems particularly promising. Given that it is often useful to combine the analysis of several types of data – ideally genetic and phenotypic – in order to detect boundaries between species, particularly in the case of complexes of species that have recently diverged, a growing number of authors are advocating the use of such an approach (Dayrat 2005; Will *et al.* 2005; Padial *et al.* 2010), rather than approaches based on the use of a single type of data, such as molecular metabarcoding (Hebert *et al.* 2003). However, in most taxonomic studies intended to be integrative, genetic and phenotypic data have been integrated in a rather "iterative" manner, which implies prioritizing these data and therefore introduces a large amount of subjectivity into this approach (Schlick-Steiner *et al.* 2010; Yeats *et al.* 2011). It should be noted that, in recent years, several authors have nevertheless endeavored to develop species delineation methods based on statistical models that make explicit assumptions about the evolutionary processes at work, in order to improve the objectivity and statistical accuracy of these delineations (Fujita *et al.* 2012; Naciri and Linder 2015).

However, it is only recently that such methods have been extended to the simultaneous analysis of genetic and non-genetic data (phenotypic or geographic, for example), thus paving the way for a truly integrative taxonomy (Guillot *et al.* 2012; Edwards and Knowles 2014; Solís-Lemus *et al.* 2015). As an example, the iBPP program, developed by Solís-Lemus *et al.* (2015), is an extension of the BPP program (Rannala and Yang 2003; Yang 2015), which allows for the analysis of DNA sequences to be combined with that of phenotypic data, whose evolution is modeled by the Brownian motion. This program allows us to test the adequacy between several species' delineation models and different types of data, analyzed simultaneously or independently, in a Bayesian framework. Thus, this method makes it possible to get rid of an *a priori* hierarchy of the different types of data and an arbitrary decision regarding the level of differentiation required for two taxa to be recognized as two distinct species.

This program has recently been used to redefine species boundaries within the group of 12 *Pseudophrys* from France, for which genetic and phenotypic data (morphology and floral scents) have been collected (Joffard 2017). Analyzed alone, genetic data only allow the recognition of four species. Analyzed in combination with morphological and/or chemical data, they allow 11 and 10 species to be distinguished, respectively, whereas the simultaneous analysis of these three types of data leads to the recognition of 10 species, the pairs *O. bilunulata*/*O. marmorata*, on the one hand, and *O. funerea*/*O. zonata*, on the other hand, being merged. These two mergers make sense since the morphological and ecological similarity of these two pairs of taxa had already been pointed out by some authors (Bournérias and Prat 2005; Tison and Foucault 2014). Moreover, it is interesting to note that morphological data have a resolving power that is equivalent, or even superior, to that of floral odor data, although the latter undeniably play a key role in speciation. Amateur orchidophiles or conservationists could therefore fully contribute to research on *Ophrys* taxonomy through the collection of particularly informative data for systematic purposes.

Nevertheless, we draw attention to two precautions to be taken when applying this integrative taxonomy approach: (1) the information provided for each species must be representative of its entire range, including, in particular, its typical locality, in order to be able to confidently associate a validated species name with the data collected; (2) it is imperative to bear in mind that the species, thus re-delineated, remain above all hypotheses that taxonomic progress (collection and inclusion of new data, for example) is likely to call into question. This approach proposes different scenarios with different numbers of species and associated likelihood values. Thus, the final choice to merge two taxa remains in the hands of the scientists adopting this approach; they must therefore explain their choice and argue for or against potential mergers. With these precautions in mind, we encourage the application of this integrative taxonomy approach to other groups of *Ophrys*, in order to resolve the problems of taxonomic confusion within the genus and improve conservation strategies for these emblematic but often threatened species.

6.5. Conclusion

In spite of the contribution of genetic tools over the past 20 years, systematics (as well as taxonomy) remains particularly problematic in the genus *Ophrys*. The increasing number of available chloroplast genomes will be accompanied in the coming months by the availability of several additional transcriptomes and, in the medium term, a high-quality reference genome for *Ophrys*. These resources will undoubtedly open a new era in the development of new genetic markers to improve knowledge of the systematics of the genus. They will also fully pave the way for the study of the genomic bases of adaptation and reproductive isolation. In parallel, their coupling with integrative taxonomy approaches, based on solid field studies integrating phenotypic (morphology, odor, etc.) and ecological (habitat variables, estimation of relative selective value) data, should enable a great leap forward in the understanding of the causes of the evolutionary radiation of *Ophrys*.

6.6. Acknowledgments

We would like to thank the following for the financial support that has contributed to this research: OSU-Oreme for the long-term observations, GDR Pollineco no. 2058 (CNRS-MTES); Nina Joffard's thesis was funded by HESAM University and its program "Paris Nouveaux Mondes". Michel Baguette and Joris Bertrand are members of the Labex TULIP (ANR-10-LABX-41); Nina Joffard and Bertrand Schatz are members of the Labex CEMEB. Michel Baguette would like to thank Pierre Devillers, Virginie M. Stevens and Daniel Tyteca for their considerable contributions.

6.7. References

Ayasse, M., Stökl, J., and Francke, W. (2011). Chemical ecology and pollinator-driven speciation in sexually deceptive orchids. *Phytochemistry*, 72(13), 1667–1677.

Baguette, M., Bertrand, J.A.M., Stevens, V.M., and Schatz, B. (2020). Why are there so many Bee-Orchid species? Adaptive radiation by intraspecific competition for mnesic pollinators. *Biol. Rev.*, 95, 1630–1663.

Bateman, R.M. (2018). Two bees or not two bees? An overview of *Ophrys* systematics. *Ber. Arbeitskrs. Heim. Orchid.*, 35(1), 5–46.

Bateman, R.M., Hollingdworth, P.M., Preston, J., Yi-Bo, L., Pridgeon, A.M., and Chase, M.W. (2003). Molecular phylogenetics and evolution of orchidineae and selected Habenariinae (Orchidaceae). *Bot. J. Linn. Soc.*, 142, 1–40.

Bateman, R.M., Bradshaw, E., Devey, D.S., Glover, B.J., Malmgren, S., Sramkó, G., and Rudall, P.J. (2011). Species arguments: Clarifying competing concepts of species delimitation in the pseudo-copulatory orchid genus *Ophrys*. *Bot. J. Linn. Soc.*, 165(4), 336–347.

Bateman, R.M., Sramkó, G., and Paun, O. (2018). Integrating restriction site-associated DNA sequencing (RAD-seq) with morphological cladistic analysis clarifies evolutionary relationships among major species groups of bee orchids. *Ann. Bot.*, 121(1), 85–105.

Bertrand, J.A.M., Gibert, A., Llauro, C., and Panaud, O. (2019). Characterization of the complete plastome of *Ophrys aveyronensis*, a Euro-Mediterranean orchid with an intriguing disjunct geographic distribution. *Mitochondrial DNA Part B*, 4(2), 3256–3257.

Bournérias, M. Prat, D. (eds) (2005). *Les Orchidées de France, Belgique et Luxembourg*, 2nd edition. Biotope Éditions, Mèze.

Breitkopf, H., Onstein, R.E., Cafasso, D., Schlüter, P.M., and Cozzolino, S. (2015). Multiple shifts to different pollinators fueled rapid diversification in sexually deceptive *Ophrys* orchids. *New Phytol.*, 207(2), 377–389.

Cortis, P., Vereecken, N.J., Schiestl, F.P., Barone Lumaga, M.R., Scrugli, A., and Cozzolino, S. (2009). Pollinator convergence and the nature of species' boundaries in sympatric Sardinian *Ophrys* (Orchidaceae). *Ann. Bot.*, 104(3), 497–506.

Danforth, B.N., Cardinal, S., Praz, C., Almeida, E.A.B., and Michez, D. (2013). The impact of molecular data on our understanding of bee phylogeny and evolution. *Annu. Rev. Entomol.*, 58, 57–78.

Dayrat, B. (2005). Towards integrative taxonomy. *Biol. J. Linn. Soc.*, 85(3), 407–415.

Delforge, P. (2016). *Orchidées d'Europe, d'Afrique du Nord et du Proche-Orient.* Delachaux & Niestlé, Lausanne.

Devey, D.S., Bateman, R.M., Fay, M.F., and Hawkins, J.A. (2008). Friends or relatives? Phylogenetics and species delimitation in the controversial European orchid genus *Ophrys*. *Ann. Bot.*, 101(3), 385–402.

Devillers, P. and Devillers-Terschuren, J. (2013). Orchidées et concepts modernes de l'espèce. *Natural. Belges*, 94(26), 61–74.

Edwards, D.L. and Knowles, L.L. (2014). Species detection and individual assignment in species delimitation: Can integrative data increase efficacy? *Proc. R. Soc. B*, 281(1777), 20132765–20132765.

Fujita, M.K., Leaché, A.D., Burbrink, F.T., McGuire, J.A., and Moritz, C. (2012). Coalescent-based species delimitation in an integrative taxonomy. *Trends Ecol. Evol.*, 27(9), 480–488.

Gervasi, D.L., Selosse, M.A., Sauve, M., Wittko, F., Vereecken, N.J., Cozzolino, S., and Schiestl, F.P. (2017). Floral scent and species divergence in a pair of sexually deceptive orchids. *Ecol. Evol.*, 7(15), 6023–6034.

Guillot, G., Renaud, S., Ledevin, R., Michaux, J., and Claude, J. (2012). A unifying model for the analysis of phenotypic, genetic, and geographic data. *Syst. Biol.*, 61(6), 897–911.

Hebert, P.D.N., Cywinska, A., Ball, S.L., and de Waard, J.R. (2003). Biological identifications through DNA barcodes. *Proc. Biol. Sci.*, 270(1512), 313–321.

Inda, L.A., Pimentel, M., and Chase, M.W. (2012). Phylogenetics of tribe Orchideae (Orchidaceae: Orchidoideae) based on combined DNA matrices: Inferences regarding timing of diversification and evolution of pollination syndromes. *Ann. Bot.*, 110(1), 71–90.

Isaac, N.J., Mallet, J., and Mace, G.M. (2004). Taxonomic inflation: Its influence on macroecology and conservation. *Trends Ecol. Evol.*, 19(9), 464–469.

IUCN France, MNHN, FCBN, SFO (2010). La liste rouge des espèces menacées de France : orchidées de France métropolitaine. Report.

Jin, W.-T., Shuiteman, A., Chase, M.W., Li, J.-W., Chung, S.-W., Hsu, T.-C., and Jin, X.H. (2017). Phylogenetics of subtribe Orchidinae s.l. (Orchidaceae; Orchidoideae) based on seven markers (plastid matK, psaB, rbcL, trnL-F, trnH-psba, and nuclear nrITS, Xdh): Implications for generic delimitation. *BMC Plant Biol.*, 17(1), 222.

Joffard, N. (2017). Diversification des orchidées méditerranéennes : niches de pollinisation, évolution des traits floraux et taxonomie intégrative. PhD Thesis, Université PSL, Paris.

Joffard, N., Massol, F., Grenié, M., Montgelard, C., and Schatz, B. (2019). Effect of pollination strategy, phylogeny and distribution on pollination niches of Euro-Mediterranean orchids. *J. Ecol.*, 107(1), 478–490.

Leitch, I.J., Kahandawala, I., Suda, J., Hanson, L., Ingrouille, M.J., Chase, M.W., and Fay, M.F. (2009). Genome size diversity in orchids: Consequences and evolution. *Ann. Bot.*, 104(3), 469–481.

Mallet, J. (2005). Hybridisation as an invasion of the genome. *Trends Ecol. Evol.*, 20(5), 229–237.

Mayr, E. (1942). *Systematics and the Origin of Species.* Columbia University Press, New York.

Naciri, Y. and Linder, H.P. (2015). Species delimitation and relationships: The dance of the seven veils. *Taxon*, 64(1), 3–16.

Nieto, A., Roberts, S.P., *et al.* (2014). *European Red List of Bees.* Office des publications de l'Union européenne, Luxembourg, 98.

Padial, J.M., Miralles, A., De la Riva, I.J., and Vences, M. (2010). The integrative future of taxonomy. *Front. Zool.*, 7, 16.

Pedersen, H.Æ. and Faurholdt, N. (2007). *Ophrys: The Bee Orchids of Europe.* Royal Botanic Gardens, London.

Piñeiro Fernández, L., Byers, K.J.R.P., Cai, J., Seedeek, K.E.M., Kellenberger, R.T., Russo, A., Qi, W., and Fournier Aquino, C. (2019) A phylogenomic analysis of the floral transcriptomes of sexually deceptive and rewarding European orchids, *Ophrys* and *Gymnadenia*. *Front. Plant Sci.*, 10, 1553.

Pridgeon, A.M., Bateman, R.M., Cox, A.V., Hapeman, J.R., and Chase, M.W. (1997). Phylogenetics of subtribe Orchidineae (Orchidoideae, Orchidaceae) based on Nuclear ITS sequences. 1. Intergeneric relationships and polyphyly of *Orchis sensu lato*. *Lindleyanna*, 12(2), 89–109.

de Queiroz, K. (2005). Different species problems and their resolution. *Bioessays*, 27(12), 1263–1269.

de Queiroz, K. (2007). Species concepts and species delimitation. *Syst. Biol.*, 56(6), 879–886.

Rannala, B. and Yang, Z. (2003). Bayes estimation of species divergence times and ancestral population sizes using DNA sequences from multiple loci. *Genetics*, 164(4), 1645–1656.

Roche, S.A., Carter, R.J., Peakall, R., Smith, L.M., Whitehead, M.R., and Linde, C.C. (2010). A narrow group of monophyletic *Tulasnella* (Tulasnellaceae) symbiont lineages are associated with multiple species of *Chiloglottis* (orchidaceae): Implications for orchid diversity. *Am. J. Bot.*, 97(8), 1313–1327.

Roma, L., Cozzolino, S., Schlüter, P.M., Scopece, G., and Cafaso, D. (2018). The complete plastid genomes of *Ophrys iricolor* and *O. sphegodes* (Orchidaceae) and comparative analyses with other orchids. *PLOS One*, 13(9), e0204174.

Schatz, B. (2017). The orchid species of Cavallo island (Lavezzi archipelago, Corsica): An amazing abundance of the protected species *Gennaria diphylla*. *Ecologia Mediterranea*, 43(2), 159–170.

Schatz, B., Geoffroy, A., Dainat, B., Bessière, J.M., Buatois, B., Hossaert-McKey, M., and Selosse, M.A. (2010). A case study of modified interactions with symbionts in a hybrid Mediterranean orchid. *Am. J. Bot.*, 97(8), 1278–1288.

Schatz, B., Sauvion, N., Kjellberg, F., and Nel, A. (2017). Plant–insect interactions: A paleontological and an evolutionary perspective. *Adv. Bot. Res.*, 81, 1–24.

Schiestl, F.P. and Ayasse, M. (2002). Do changes in floral odor cause speciation in sexually deceptive orchids? *Plant Syst. Evol.*, 234(1-4), 111–119.

Schlick-Steiner, B.C., Steiner, F.M., Seifert, B., Stauffer, C., Christian, E., and Crozier, R.H. (2010). Integrative taxonomy: A multisource approach to exploring biodiversity. *Annu. Rev. Entomol.*, 55, 421–438.

Schlüter, P.M., Kohl, G., Stuessy, T.F., and Paulus, H.F. (2007). A screen of low-copy nuclear genes reveals the LFY gene as phylogenetically informative in closely related species of orchids (*Ophrys*). *Taxon*, 56(2), 493–504.

Scopece, G., Musacchio, A., Widmer, A., and Cozzolino, S. (2007). Patterns of reproductive isolation in Mediterranean deceptive orchids. *Evolution*, 61(11), 2623–2642.

Sedeek, K.E.M., Qi, W., Schauer, M.A., Gupta, A.K., Poveda, L., Xu, A., Liu, Z., Grossniklaus, U., Schiestl, F.P., and Schlüter, P.M. (2013). Transcriptome and proteome data reveal candidate genes for pollinator attraction in sexually deceptive orchids. *PLOS One*, 8(5), e64621.

Sedeek, K.E.M., Scopece, G., Staedler, Y.M., Schönenberger, J., Cozzolino, S., Schiestl, F.P., and Schlüter, P.M. (2014). Genic rather than genome-wide differences between sexually deceptive *Ophrys* orchids with different pollinators. *Mol. Ecol.*, 23(24), 6192–6205.

Solís-Lemus, C., Knowles, L.L., and Ané, C. (2015). Bayesian species delimitation combining multiple genes and traits in a unified framework. *Evolution*, 69(2), 492–507.

Soliva, M., Kocyan, A., and Widmer, A. (2001). Molecular phylogenetics of the sexually deceptive orchid genus *Ophrys* (Orchidaceae) based on nuclear and chloroplast DNA sequences. *Mol. Phylogenet. Evol.*, 20(1), 78–88.

Tison, J. and de Foucault, B. (2014). *Flora gallica : flore de France*. Biotope Éditions, Mèze.

Triponez, Y., Arrigo, N., Pellissier, L., Schatz, B., and Alvarez, N. (2013). Morphological, ecological and genetic aspects associated with endemism in the Fly Orchid group. *Mol. Ecol.*, 22(5), 1431–1446.

Tyteca, D. and Baguette, M. (2017). *Ophrys* (Orchidaceae) systematics – When molecular phylogenetics, morphology and biology reconcile. *Ber. Arbeitskrs. Heim. Orchid.*, 34(1), 37–103.

Vereecken, N.J., Cozzolino, S., and Schiestl, F.P. (2010). Hybrid floral scent novelty drives pollinator shift in sexually deceptive orchids. *BMC Evol. Biol.*, 10, 103.

Vereecken, N.J., Streinzer, M., Ayasse, M., Spaethe, J., Paulus, H.F., Stökl, J., and Schiestl, F.P. (2011). Integrating past and present studies on *Ophrys* pollination: A comment on Bradshaw *et al. Bot. J. Linn. Soc.*, 165, 329–335.

Will, K.W., Mishler, B.D., and Wheeler, Q.D. (2005). The perils of DNA barcoding and the need for integrative taxonomy. *Syst. Biol.*, 54(5), 844–851.

Wright, S. (1932). The roles of mutation, inbreeding, crossbreeding, and selection in evolution. *Proceedings of the Sixth International Congress of Genetics*, 1, 356–366.

Yang, Z. (2015). The BPP program for species tree estimation and species delimitation. *Curr. Zool.*, 61(5), 854–865.

Yeates, D.K., Seago, A., Nelson, L., Cameron, S.L., Joseph, L., and Trueman, J.W.H. (2011). Integrative taxonomy, or iterative taxonomy? *Syst. Entomol.*, 36(2), 209–217.

7

Exploration and Origins of Biodiversity in Madagascar: The Message of Ferns

7.1. Introduction

Madagascar is a continental island in the Indian Ocean, renowned for both its extraordinary biodiversity and the threats it faces. This justifies its inclusion among the biodiversity hotspots (Myers *et al.* 2000), and it has even been described as the "hottest biodiversity hotspot" (Ganzhorn *et al.* 2001). For example, 81% of vascular plants (MadCat 2019), 52% of birds and 86% of invertebrates are endemic, and this proportion even reaches 100% for terrestrial gastropods and non-flying mammals (Goodman and Benstead 2005).

These characteristics make it a particularly interesting field of study for the regional and international scientific community, and in particular for the researchers of the *Muséum national d'histoire naturelle* (MNHN; French National Museum of Natural History), who have been conducting research there for nearly 350 years. It was, indeed, as early as 1648 that Etienne de Flacourt, the appointed governor of Madagascar, constituted a small herbarium of a few dozen Malagasy plants; it is the oldest tropical herbarium in the world. Among these plants was the Madagascar Periwinkle, one of the emblematic plants of Madagascar known for its anti-cancerous active ingredients. The relationship between Madagascar and the Herbarium of the King's Garden (which became the French National Museum of Natural History in 1793) thus started with these plants, which have been preserved

Chapter written by Germinal ROUHAN and Myriam GAUDEUL.

since then at the French National Herbarium. Nowadays, the specimens of plants from Madagascar in the herbarium of this museum constitute one of the most important, if not the richest, collections in the world, thanks to multiple expeditions and works led by renowned botanists such as Henri Perrier de la Bâthie (1873–1958) and Henri Humbert (1887–1967).

In spite of the sampling effort, the richness of the collections and the published works, the biological diversity of Madagascar remains poorly known, with potentially more than 2,000 species of vascular plants yet to be described (MadCat 2019). Similarly, knowledge of the evolutionary processes at the origin of this diversity was almost non-existent until recently, with only a few phylogenetic studies available. We are therefore developing two main lines of research in Madagascar: (1) the description of the specific diversity with integrative taxonomic approaches; (2) the study of the origin of endemic lineages with biogeographic approaches. We are particularly interested in assessing the relative contribution of vicarious processes generated by plate tectonics, long-distance transoceanic dispersions and *in situ* species diversification. Our approaches are empirical through the analysis of molecular phylogenies of plants taken as study models and through the examination of whether a trend emerges, which is common to several groups of plants, in the inferred biogeographic processes that led to the diversity observed today for these groups in Madagascar.

7.2. Madagascar: a complex biogeographical context

7.2.1. *An insular continental territory that is not so isolated*

Regardless of the organisms of interest, studying the origin and evolutionary history of Malagasy lineages is a challenge, given the complex history of this island landmass that is surrounded by several archipelagos and the fact that it has been isolated for a long time, even though it is only separated from the East African coast by less than 500 kilometers (Figure 7.1).

Thus, Madagascar must be studied in the broader context of the western Indian Ocean, involving both continental and oceanic islands (Agnarsson and Kuntner 2012). Madagascar is also part of the geological context of the Gondwanaland supercontinent, which began to fragment 155–118 million years ago with the separation of an America/Africa block from the Madagascar/India/Australia/Antarctica block. Rapidly thereafter, America and Africa separated from each other 110 million years ago, before India and the Seychelles drifted together northward, leaving Madagascar isolated for 118–84 million years; followed by the separation of the Seychelles Plateau (100–95 million years) (Seward *et al.* 2004) and finally the collision of India with Asia 50 million years ago. At its present geographical position, Madagascar is surrounded by two volcanic archipelagos (Comoros and Mascarene Islands), which began to emerge, respectively, 15 (Emerick and Duncan 1982; Nougier *et al.* 1986) and 8 million years ago (McDougall and Chamalaun 1969).

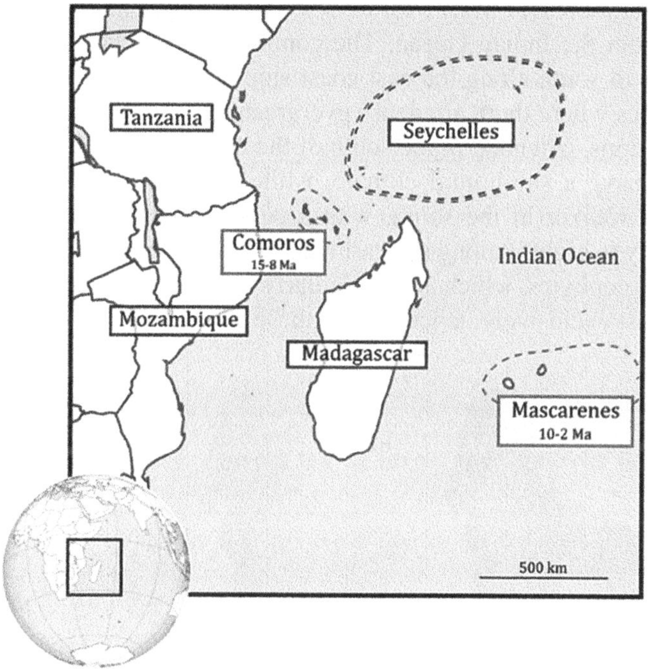

Figure 7.1. *Map of the western Indian Ocean, locating Madagascar and the neighboring islands (the territories of continental origin appear in blue and those of volcanic origin in red). For a color version of this figure, see www.iste.co.uk/grandcolas/systematics.zip*

Studies on Madagascar species have shown that the origin of diversity is related to continental drift for some organisms, while other more recent research shows a greater role for long-distance dispersion during the Cenozoic Era (Yoder and Nowak 2006; Renner *et al.* 2010; Warren *et al.* 2010; Agnarsson and Kuntner 2012; Buerki *et al.* 2013; Bacon *et al.* 2016). The paramount importance of dispersions is even more evident in the origin of the lineages present on the recent volcanic islands surrounding Madagascar.

7.2.2. *Gradients, ecosystem diversity and biodiversity*

The ecosystems present in Madagascar are varied and contrasting. The island extends over 1,600 kilometers, mainly in the intertropical zone. However, the diversity of ecosystems is mainly related to a moisture gradient that increases from west to east as a result of moisture-laden winds blowing from the Indian Ocean. The continuous mountain range extending from north to south along the east coast stops the winds, generating rainfall on the slopes where there are dense evergreen rainforests, some of which are rare formations, extended in the south of the tropics (Figure 7.2). The central highlands have a sub-humid climate, while the west is dry. The moisture gradient is weaker in the south, with a sub-arid climate in the southwest. Biodiversity is highest along the east coast, in dense humid forests. We study ferns and lycophytes, which are particularly representative of this ecosystem, being almost exclusively dependent on these environments.

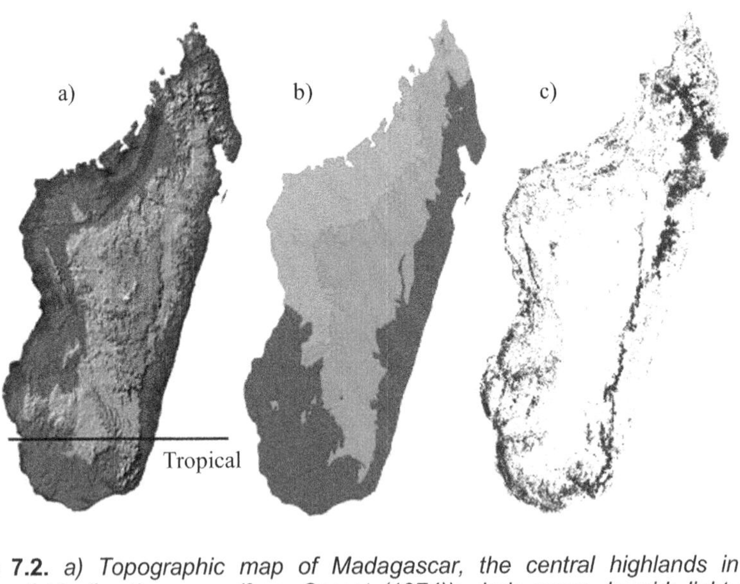

Figure 7.2. *a) Topographic map of Madagascar, the central highlands in gray. b) Map of bioclimatic zones (from Cornet (1974)): dark green, humid; light green, sub-humid; yellow, dry; red, sub-arid. c) Map of current forests (Vieilledent* et al. *2018). For a color version of this figure, see www.iste.co.uk/grandcolas/systematics.zip*

Figure 7.3. *Diversity of Madagascar's ferns and lycophytes, illustrated by 10 species that are representative of the main lineages. For a color version of this figure, see www.iste.co.uk/grandcolas/systematics.zip*

COMMENT ON FIGURE 7.3. – *a)* Rumohra linearisquamata *Rakotondr. b)* Marsilea quadrifolia *L. c)* Hymenophyllum poolii *Baker. d)* Equisetum *sp.*

e) Phlegmariurus trigonus *(C.Chr.) A.R.Field & Bostock. f)* Cyathea boivinii *Mett. ex Kuhn var.* andringitrensis *Janssen and Rakotondr. g)* Cheiroglossa palmata *(L.) C. Presl. h)* Selaginella *sp. i)* Elaphoglossum marojejyense *Tardieu. j)* Psilotum nudum *(L.) P. Beauv.*

7.3. Ferns and lycophytes: an ideal model for the biogeography of Madagascar

Ferns and lycophytes (still called "pteridophytes", even if they do not form a monophyletic group together) are the two lineages of vascular plants with spores. We use them as biological models in biogeographic studies, as they present several traits that allow us to test different hypotheses about the origin of the lineages present in Madagascar.

The exploration and description of the diversity of species of these groups has been the subject of numerous field missions and research projects, and it is expected that about 700 species will be included in the new "Flora of the Pteridophytes of Madagascar", coordinated at the MNHN by F. Rakotondrainibe (Figure 7.3). This diversity is, on the one hand, very high, since it represents more than 6% of the world's diversity, whereas Madagascar's surface area only represents 1% of the land surface. On the other hand, the diversity of pteridophytes is also particularly original, with an endemism rate of more than 50%. As in other territories, however, endemism is not as high as in flowering plants and other previously mentioned groups, probably because fern spores show a strong potential for dispersal, which favors the arrival of species, maintains genetic exchanges and finally slows down the speciation process (Smith 1972). Dispersion of the tiny spores (these are isolated cells) is largely ensured by winds, although other dispersion vectors have also been documented, such as bats (Sugita *et al.* 2013) or birds (Lewis *et al.* 2014).

7.4. Origins of the lineages of ferns in Madagascar

7.4.1. *Multiple long-distance dispersions*

Spore dispersion capacity, associated with the relatively recent origin of a large number of fern lineages since the Cretaceous (< 150 Ma) (Schuettpelz and Pryer 2009; Rothfels *et al.* 2015), suggests a predominant role of long-distance dispersion (compared to the Gondwanaland vicariance) in explaining the origin of the lineages present in Madagascar. In particular, and despite the remoteness of these two geographical areas, connections between Neotropics and Madagascar had already been suggested – first on the basis of floristic affinities (Moran and Smith 2001) and then using molecular phylogenies on the genera *Elaphoglossum* Schott ex J. Sm. (Rouhan *et al.* 2004; Vasco *et al.* 2009), *Lomariopsis* Fée (Rouhan *et al.* 2007), *Stenogrammitis* Labiak (Labiak *et al.* 2010) or *Leucotrichum* Labiak (Rouhan *et al.* 2012). Dated phylogenies have led to similar inferences of transatlantic dispersions in the genera *Alsophila* R.Br. (Janssen *et al.* 2008) and *Parapolystichum* (Keyserl.) Ching (Labiak *et al.* 2014).

Our more recent comparative biogeographic research has targeted about 20 other genera, in order to situate 84 endemic species of Madagascar in global spatiotemporal frameworks, using dated molecular phylogenies with sampling that is as exhaustive as possible. We have thus shown that the endemic fern lineages are all less than 35 million years old and the majority of them are even less than 10 million years old. Knowing that Madagascar has been isolated for about 90 million years, the vicariance hypothesis is difficult to support, and long-distance dispersion is the preferred hypothesis to explain the origin of the endemic diversity of ferns in Madagascar.

7.4.2. *The Neotropics: a non-exclusive but preponderant role*

The reconstruction of the ancestral distribution areas of the 20 or so generic lineages involving species that are endemic to Madagascar has confirmed the close relationship between the Neotropics and Madagascar, with the inference of about 10 independent dispersion events. Apart from these, our results showed six to seven dispersions of Asian origin, and three dispersions from the nearest continent, Africa (Bauret *et al.* 2017a, 2017b, 2018).

7.4.3. *Africa: a truly minimal role or an underestimated role?*

We only found a few dispersions of African origin, in grammitid ferns (Bauret *et al.* 2017a) and the genus *Phlegmariurus* Holub (Bauret *et al.* 2018). A few other studies have inferred the same type of dispersion in other fern groups, but never in large numbers (in the genera *Platycerium* Desv. and *Alsophila*) (Kreier and Schneider 2006; Janssen *et al.* 2008). Conversely, dispersion from Africa is a very strong trend for other plant groups (Agnarsson and Kuntner 2012).

Africa's minimal role in the origin of the Malagasy fern lineages is surprising, considering its greater geographical proximity and its very large surface compared to Madagascar. Africa's apparently less diverse source could be due to less speciation on this continent, since the current diversity of ferns does not exceed 800 species (compared to 3,500 in South America and 4,500 in Southeast Asia). This lower diversity may also be related to a decline in African forests during the Cenozoic Era, associated with mass extinctions (Kornas 1993; Morley 2000; Linder *et al.* 2014). Ultimately, Africa's reduced forest area may not have constituted a sufficient reservoir of species that could then spread to Madagascar. Nevertheless, we cannot exclude the fact that extinctions in Africa, through the masking of African lineages in the resulting phylogenies, may influence the inference towards a greater number of dispersions from the Neotropics, when at least some of these could in fact have originated on the African continent (Bauret *et al.* 2017a). The fossil record of ferns in Africa is too patchy to test this hypothesis.

7.5. The example of *Rumohra*: dispersions to Madagascar and around the world

The genus *Rumohra* Raddi (Dryopteridaceae), although poorly diversified with only seven species, constitutes a good model for the study of the biogeography and origin of lineages in Madagascar. Indeed, it has three species that are endemic to Madagascar, but is also widely distributed globally with three other species in South America and one that is largely distributed in the Southern Hemisphere.

Based on an exhaustive sampling of the species, including 28 samples of the most widely distributed species (*R. adiantiformis* (G. Forst) Ching), we found a phylogenetic pattern that demonstrates the existence of at least six lineages of *R. adiantiformis*. This morphological species actually represents several cryptic species, each corresponding to a distinct geographical region. The diversity of Malagasy species would most probably result from a long-distance dispersion event from the Neotropics, followed by *in situ* diversification.

Our sampling also allowed us to propose a global biogeographic scenario for the genus (Figure 7.4): while the genus diverged from the sister genus *Megalastrum* Holttum more than 46 million years ago, its diversification only began 11 million years ago, and its biogeographic history is marked by seven long-distance dispersion events, three of which started in the Neotropics and spread to South Africa, the western Indian Ocean and southern South America. The Australasian lineage would result from long-distance dispersion from South America, again illustrating the great dispersion capacity of these organisms.

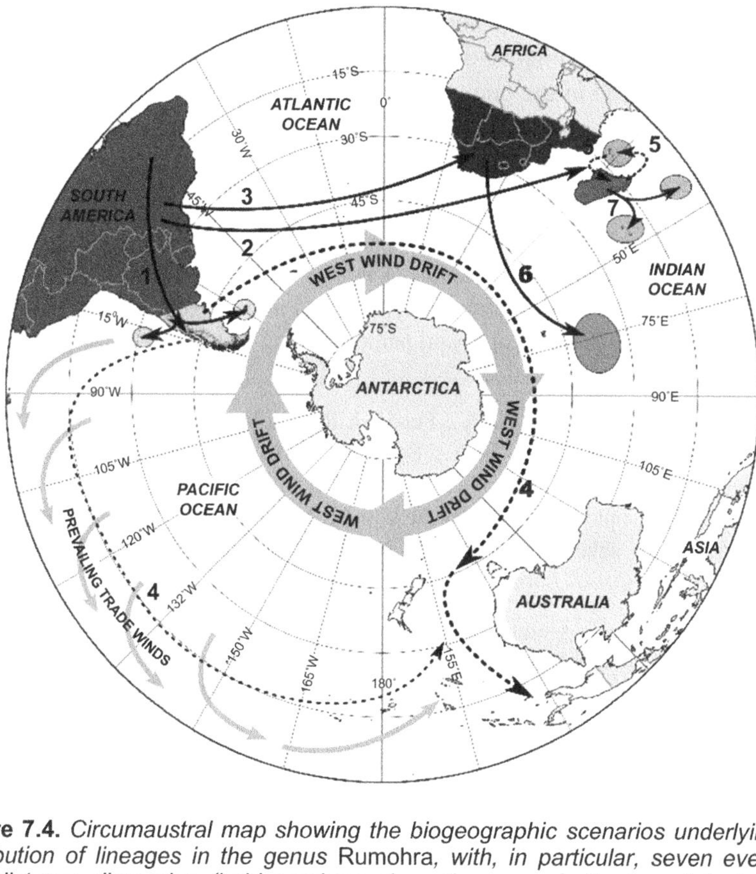

Figure 7.4. *Circumaustral map showing the biogeographic scenarios underlying the distribution of lineages in the genus* Rumohra, *with, in particular, seven events of long-distance dispersion (bold numbers along the arrows); the two dotted arrows indicate two possible scenarios for dispersion from South America to Australasia (from Bauret* et al. *(2017b)). For a color version of this figure, see www.iste.co.uk/grandcolas/systematics.zip*

7.6. Conclusion

Our studies on the processes behind the diversity of Madagascar's ferns and lycophytes have shown that close biogeographic connections seem to exist between Madagascar and other regions of the Southern Hemisphere, in particular South America, resulting from long-distance transoceanic dispersions. Thus, contrary to the hypothesis still frequently put forward, Africa was not the "reservoir of diversity" for Madagascar. Moreover, contrary to another widespread idea, Madagascar's endemic diversity would not – or not only – result from *in situ* diversifications: these would be rather rare for ferns, even if this tendency remains to be confirmed. Conversely, there seem to be more intense diversifications in lycophytes, according to the first results obtained in the genus *Phlegmariurus* (Bauret *et al.* 2018).

7.7. References

Bauret, L., Gaudeul, M., Sundue, M.A., Parris, B.S., Ranker, T.A., Rakotondrainibe, F., Hennequin, S., Ranaivo, J., Selosse, M.-A., and Rouhan, G. (2017a). Madagascar sheds new light on the molecular systematics and biogeography of grammitid ferns, new unexpected lineages and numerous long-distance dispersal events. *Mol. Phylogenet. Evol.*, 111, 1–17.

Bauret, L., Rouhan, G., Hirai, R.Y., Perrie, L.R., Prado, J., Salino, A., Senterre, B., Shepherd, L.D., Sundue, M.A., Selosse, M.-A., and Gaudeul, M. (2017b). Molecular data, based on an exhaustive species sampling of the fern genus *Rumohra* (Dryopteridaceae), reveal a biogeographic history mostly shaped by dispersal and several cryptic species in the widely distributed *Rumohra adiantiformis*. *Bot. J. Linn. Soc.*, 185, 463–481.

Bauret, L., Field, A.R., Gaudeul, M., Selosse, M.-A., and Rouhan, G. (2018). First insights on the biogeographical history of *Phlegmariurus* (Lycopodiaceae), with a focus on Madagascar. *Mol. Phylogenet. Evol.*, 127, 488–501.

Cornet, A. (1974). Essai de cartographie bioclimatique à Madagascar. Notice explicative. Essay, ORSTOM, Paris.

Janssen, T., Bystriakova, N., Rakotondrainibe, F., Coomes, D., Labat, J.N., and Schneider, H. (2008). Neoendemism in Madagascan scaly tree ferns results from recent, coincident diversification bursts. *Evolution*, 62(8), 1876–1889.

Kornas, J. (1993). The significance of historical factors and ecological preference in the distribution of African Pteridophytes. *J. Biogeogr.*, 20, 281–286.

Kreier, H.P. and Schneider, H. (2006). Phylogeny and biogeography of the staghorn fern genus *Platycerium* (Polypodiaceae, Polypodiidae). *Am. J. Bot.*, 93, 217–225.

Labiak, P.H., Sundue, M., and Rouhan, G. (2010). Molecular phylogeny, character evolution, and biogeography of the grammitid fern genus *Lellingeria* (Polypodiaceae). *Am. J. Bot.*, 97, 1354–1364.

Labiak, P.H., Sundue, M., Rouhan, G., Hanks, J.G., Mickel, J.T., and Moran, R.C. (2014). Phylogeny and historical biogeography of the lastreopsid ferns (Dryopteridaceae). *Am. J. Bot.*, 101, 1207–1228.

Lewis, L.R., Behling, E., Gousse, H., Qian, E., Elphick, C.S., Lamarre, J.-F., Bêty, J., Liebezeit, J., Rozzi, R., and Goffinet, B. (2014). First evidence of bryophyte diaspores in the plumage of transequatorial migrant birds. *PeerJ*, 2, e424.

Linder, H.P., Pennington, T., and Schneider, H. (2014). The evolution of African plant diversity. *Front. Ecol. Evol.*, 2, 1–14.

MadCat (2019). Madagascar Catalogue, Catalogue of the Vascular Plants of Madagascar. Missouri Botanical Garden, Saint-Louis, United States, Antananarivo, Madagascar [Online]. Available: http://www.tropicos.org/Project/Madagascar [Accessed 2 October 2019].

Moran, R.C. and Smith, A.R. (2001). Phytogeographic relationships between neotropical and African-Madagascan pteridophytes. *Brittonia*, 53, 304–351.

Morley, R.J. (2000). *Origin and Evolution of Tropical Rain Forests*. John Wiley & Sons, New York.

Rothfels, C.J., Li, F.W., Sigel, E.M., Huiet, L., Larsson, A., Burge, D.O., Ruhsam, M., Deyholos, M., Soltis, D.E., Stewart Jr, C.N., Shaw, S.W., Pokorny, L., Chen, T., de Pamphilis, C., De Gironimo, L., Chen, L., Wei, X., Sun, X., Korall, P., Stevenson D.W., Graham, S.W., Wong, G.K., Pryer, K.M. (2015). The evolutionary history of ferns inferred from 25 low-copy nuclear genes. *Am. J. Bot.*, 102, 1–19.

Rouhan, G., Dubuisson, J.-Y., Rakotondrainibe, F., Motley, T.J., Mickel, J.T., Labat, J.N., and Moran, R.C. (2004). Molecular phylogeny of the fern genus *Elaphoglossum* (Elaphoglossaceae) based on chloroplast non-coding DNA sequences, contributions of species from the Indian Ocean area. *Mol. Phylogenet. Evol.*, 33, 745–763.

Rouhan, G., Hanks, J.G., McClelland, D., and Moran, R.C. (2007). Preliminary phylogenetic analysis of the fern genus *Lomariopsis* (Lomariopsidaceae). *Brittonia*, 59, 115–128.

Rouhan, G., Labiak, P.H., Randrianjohany, E., and Rakotondrainibe, F. (2012). Not so neotropical after all, the Grammitid fern genus *Leucotrichum* (Polypodiaceae) is also paleotropical, as revealed by a new species from Madagascar. *Syst. Bot.*, 37, 331–338.

Schuettpelz, E. and Pryer, K.M. (2009). Evidence for a Cenozoic radiation of ferns in an angiosperm-dominated canopy. *Proc. Natl. Acad. Sci.*, 106, 11200–11205.

Smith, A.R. (1972). Comparison of fern and flowering plant distributions with some evolutionary interpretations for ferns. *Biotropica*, 4, 4–9.

Sugita, N., Ootsuki, R., Fujita, T., Murakami, N., and Ueda, K. (2013). Possible spore dispersal of a bird-nest fern *Asplenium setoi* by bonin flying foxes *Pteropus pselaphon*. *Mammal Study*, 38, 225–229.

Vasco, A., Moran, R.C., and Rouhan, G. (2009). Circumscription and phylogeny of the *Elaphoglossum ciliatum* group (*E.* sect. *Lepidoglossa*, Dryopteridaceae) based on cpDNA sequences. *Taxon*, 58, 825–834.

Vieilledent, G., Grinand, C., Rakotomalala, F.A., Ranaivosoa, R., Rakotoarijaona, J.R., Allnutt, T.F., and Achard, F. (2018). Combining global tree cover loss data with historical national forest cover maps to look at six decades of deforestation and forest fragmentation in Madagascar. *Biol. Cons.*, 222, 189–197.

8

Mediterranean and Atlantic Algae, a Fraternal Relationship?

The Mediterranean Sea is the cradle of a great diversity of marine organisms. Among these, approximately 1,100 species of algae are currently recognized as belonging to the three lineages of marine macroalgae (green, red and brown algae), with contrasting evolutionary histories.

The specific diversity of the Mediterranean represents nearly 13% of the world's algae diversity. However, the origin of this diversity and its relationship with the adjacent Atlantic flora is still rather poorly understood.

We have used tools of molecular systematics (barcode marker CO1) to examine the relationship between Atlantic and Mediterranean algae species. The phylogeographic models of the Mediterranean marine flora revealed by the molecular systematics tools suggest a significant flow of species from the Atlantic Ocean to the Mediterranean Sea, through the Strait of Gibraltar. However, a few Mediterranean lineages without close allies detected throughout the world have been highlighted. The implications of these results will be discussed in terms of conservation and dynamics of biodiversity, in the current context of global change. Finally, the possible causes of the underestimation of algal diversity in the Mediterranean Sea will be examined in light of the history of the phycology of this region.

Chapter written by Line LE GALL, Delphine GEY and Florence ROUSSEAU.

8.1. Introduction

8.1.1. *Seaweeds*

Seaweeds mainly colonize and grow along rocky coastlines, where they contribute significantly to the coastal ecosystem. They play an important role as primary producers by carrying out photosynthesis and also by constituting habitats for other species which find refuge there; certain species of algae such as kelp or coralline algae are qualified as "engineers species". In temperate regions of the world, they occupy a crucial place in rocky coastal ecosystems, where they often have very dense populations. They sometimes dominate the underwater seascape in the form of forests or meadows, whose photosynthetic activity ensures one of the highest productivities recorded: a coastal zone dominated by algae thus produces more than a terrestrial forest or a cultivated field.

8.1.2. *The systematics of algae*

The systematics of algae can be traced back to the modern system of binominal nomenclature designed by the Swedish naturalist Carl Linnaeus, which has been used as a reference since the publication of *Species Plantarum* in 1753. At the time, the few recognized types of algae were placed in the subdivision of algae of the Cryptogams class, which also included ferns, mosses and fungi. The Cryptogams class juxtaposed the 23 classes of conical and flowering plants within the plant kingdom. The algae were distributed on the basis of their structures and morphological organization within three genera: *Fucus* for the tough algae, *Ulva* for membranous algae and *Conferva* for those with a filamentous appearance. This initial classification largely underestimated the diversity of the algae and, as a result, was subjected to numerous successive reshuffles. Jean Vincent Félix Lamouroux (1779–1825), Professor of Natural History at the University of Caen in France, presented, in his *Essay on the Genera of the Family of Non-Articulated Thalassiophytes* (1813), a classification of algae based on their green, brown or red color and thus laid the foundation for the distinction of the three main lineages of large algae that are still recognized today. This classification scheme was approved by Carl Adolph Agardh (1785–1859), who adopted it in his *Synopsis algarum Scandinaviae* (1817) and recognized the importance of colors in establishing a natural classification system in his *Species algarum* (1821). In 1836, the botanist and physician William Henry Harvey (1811–1866) formalized the use of colors to divide large seaweeds into green, brown and red algae, as currently recognized to a few approximations.

Since this founding work, no less than 6,000 red seaweeds, 2,000 brown seaweeds and 2,000 green seaweeds have been described (Guiry *et al.* 2014; Guiry and Guiry 2020). Since the dawn of the 21st century, the advent of PCR and DNA sequencing has made it possible to have molecular traits whose comparative analysis allows phylogenetic reconstructions. Numerous works attempt to answer the question of the origin of the diversity of Eukaryotes (e.g. Adl *et al.* (2019)) and that of the scattering of photosynthetic organisms in different lineages of the tree of life (e.g. Ponce-Toledo *et al.* (2017)), which is explained by the theory of endosymbiosis, popularized by the work of Lynn Margulis. Assessing the diversity of algal species is a difficult task because their identification on the basis of morpho-anatomical traits is often difficult due to phenomena of convergence or phenotypic plasticity, as well as complex life cycles, which are sometimes heteromorphic. The use of DNA sequences to delineate and identify algae species (e.g. Saunders 2005) has gained momentum since the launch of the international DNA barcoding project (Hebert *et al.* 2003). However, the library of available sequences (Le Gall *et al.* 2017) is far from covering the extent of the diversity of algae described and certainly even further from covering the actual diversity of algae, considering that it is now estimated that only 20% of the species present on Earth have been described (Mora *et al.* 2011).

8.1.3. *Algae distribution on a global scale*

The vast majority of marine organisms are at their optimum in terms of species diversity between the tropics (Tittensor *et al.* 2010). However, seaweeds escape this biodiversity pattern with an optimum diversity that can be observed in temperate regions (Kerswell 2006). On a finer taxonomic scale (Keith *et al.* 2014), green algae have their optimum diversity at lower latitudes than red and brown algae, respectively. The main argument put forward to explain this deviation from the general pattern lies in the biotic interactions between algae and corals; the latter dominate the tropical zones, resulting in lower algal diversity. However, recent work on species hitherto considered as pantropical suggests that tropical species diversity has been grossly underestimated, as is the case for the genus of brown algae *Lobophora* (Schultz *et al.* 2015; Vieira *et al.* 2016) and one of red algae

Portieria (Leliaert *et al.* 2018), each of which has increased from a handful to about 100 species. While these recent discoveries may qualify the understanding of latitudinal gradients in algae diversity, the fact remains that the European Atlantic and Mediterranean coasts are hotspots for algal biodiversity and are home to 700 (Bunker *et al.* 2017) and 1,100 (Boudouresque 2004; Rodríguez-Prieto 2013) macro-algae species, respectively.

Figure 8.1. *The underwater vegetation of the Atlantic is generally composed of* Fucus *in the intertidal zone, and kelp (left) in the subtidal zone. In the Mediterranean, forests of cystoseira in the upper zone and deeper coralligenous reefs (right) are emblematic elements of the vegetation. For a color version of this figure, see www.iste.co.uk/grandcolas/systematics.zip*

8.1.4. *Seaweeds on the Atlantic and Mediterranean coasts*

The formalization of knowledge about algae dates back to antiquity. Pliny the Elder refers to it in his book *The Natural History*, where he mentions the property of certain algae, and particularly the dyeing property of a Mediterranean species (probably *Rytiphlaea tinctoria*). Moreover, he indicates that the algae populations of the Mediterranean are less than those of the ocean. However, in this context, it is not easy to determine whether he is referring to the size of the organisms or their diversity. It was only at the beginning of the 19th century, in line with the work undertaken by Buffon on the geographical distribution of vertebrate animals, as well as by Jean-Baptiste de Lamarck and Augustin Pyrame de Candolle on the flora of France (1806), that Lamouroux (1828) became interested in the geography of algae and estimated that a third of the 590 species of algae that he hoarded in his French herbarium were found in both the Atlantic and the Mediterranean. Among the numerous summaries of knowledge on Mediterranean marine biodiversity (e.g. Coll *et al.* 2010; Stambler 2012), the one carried out by Charles-François Boudouresque (2004) devotes particular attention to algae. Mediterranean algae account for 13% of the global specific diversity of algae (Giaconne 1974) and their endemicity rate in the Mediterranean is in the order of 20%, which is of the same order of magnitude as the rates calculated for marine invertebrates and fish (Fredj *et al.* 1992). Since these estimates, lists of Mediterranean species have been established for green (Benhissoune *et al.* 2001), brown (Benhissoune *et al.* 2002) and an order of red algae, Ceramiales (Gómez Garreta *et al.* 2001). However, work on molecular systematics shows that knowledge of Mediterranean algae is still very incomplete (e.g. Manghisi *et al.* 2014; Vergés *et al.* 2014a; Peña *et al.* 2015; Le Gall *et al.* 2018; De Jode *et al.* 2019; Rindi *et al.* 2019).

8.1.5. *Challenge of the study*

Jean Feldmann (1937) emphasizes that through its biogeography, "France is, of all European countries, the best placed for research in marine phycology. The great majority of the species of temperate Europe can be found on its western coasts of the English Channel and the ocean; whereas on its Mediterranean coasts, the marine flora, which is very different and almost subtropical, presents a large number of interesting species, and the particular ecological conditions of this sea are worthy of attracting phyto-oceanographers who are able to compare its vegetation with that of the neighboring ocean". In order to explain the origin of the diversity of the Mediterranean algal flora, Feldmann proposes to break down this flora into five elements: Cosmopolitan element, pantropical element, Indo-Pacific element, Atlantic element (boreal and tropical) and endemic element. Today, it is appropriate to add the lessepsian element in order to consider the species introduced into the Mediterranean via the Suez Canal. According to Feldmann, the Atlantic element is by far the richest in species, and this assertion is confirmed by the 47% estimate of the Atlantic element in the algal flora of Sicily (Giaconne and Geraci 1989). In fact, following the Messinian salinity crisis, the Zanklean transgression (about 5.33 million years ago) allowed a very rapid re-watering of the Mediterranean and would have brought a large number of species from the Atlantic to the Mediterranean. After that transgression, however, it is generally accepted that the Strait of Gibraltar constitutes an important biogeographic barrier that limits trade between the Ocean and the Mediterranean (Borsa *et al.* 1997). However, it appears that the transition between the Atlantic and the

Mediterranean leads to numerous patterns of diversity, which, to date, remain poorly understood (Patarnello *et al.* 2007).

In order to better assess the biogeographic relationships between Mediterranean and Atlantic algae, we focused on two genera of red algae of the order Rhodymeniales, *Lomentaria* (Lomentariaceae) and *Champia* (Champiaceae), whose distribution includes the Atlantic and the Mediterranean, and we used the tools of molecular systematics to study their potential biogeographic separation.

8.2. Materials and methods

8.2.1. *Sampling strategy*

Figure 8.2. *Algae can be collected during low tides on the Atlantic coast, or by diving, to collect species living below the tidal sway zone, which constitutes the bulk of the flora in the Mediterranean. The strategy is to sample all the morphs and habitats of all species of algae for each site surveyed. For a color version of this figure, see www.iste.co.uk/grandcolas/systematics.zip*

For the past 12 years, we have been carrying out numerous field missions in France, Europe and around the world to make an inventory of the diversity of benthic organisms and, in particular, seaweeds. During these missions, for each site prospected, we strive to collect at least one specimen of each morph per habitat, whether that be at low tide or by diving. After collection, the specimens are pressed on paper and dried with newspaper in order to make herbarium vouchers that can be preserved for several centuries. Whenever possible, a fragment of an individual preserved in a herbarium is collected and preserved in silica gel, which allows rapid tissue dehydration and thus good DNA preservation. Once dry, the tissue is stored in tubes at -20°C. The collection information (date, location, habitat, depth, collectors, means of collection, etc.) is recorded in the database.

Specimen identifier	Taxon	Region and date of collection	Collectors
Champiaceae			
LLG1603	*Champia* sp1	Ille-et-Vilaine, Brittany, 15/07/08	L. Le Gall, J. Utge, F. Rousseau
LLG1740	*Champia* sp1	Ille-et-Vilaine, Brittany, 16/07/08	L. Le Gall, J. Utge, F. Rousseau, L.E. Hochart
LLG1783	*Champia* sp1	Ille-et-Vilaine, Brittany, 16/07/08	L. Le Gall, J. Utge, F. Rousseau, L.E. Hochart
LLG1842	*Champia* sp1	Channel, Normandy, 26/08/08	L. Le Gall, I. Rauss, B. Martin, Y. Turpin, T. Abiven
LLG1912	*Champia* sp1	Channel, Normandy, 26/08/08	L. Le Gall, I. Rauss, B. Martin, Y. Turpin, T. Abiven
LLG1985	*Champia* sp1	Channel, Normandy, 27/08/08	L. Le Gall, I. Rauss, B. Martin, Y. Turpin, T. Abiven, J. Utge
LLG2916	*Champia* sp1	Finistère, Brittany, 06/08/09	J. Utge, Y. Turpin, A. Besnier
LLG2987	*Champia* sp1	Finistère, Brittany, 12/08/09	L. Le Gall, J. Utge, Y. Turpin
LLG2347	*Champia* sp2	Var, Provence-Alpes-Côte d'Azur, 21/10/08	L. Le Gall, F. Rousseau
LLG6618	*Champia* sp2	Corsica, 22/05/19	E. Vassard, A. Krygelmans, B. Gouillieux, L. Le Gall
LLG6623	*Champia* sp2	Corsica, 22/05/19	E. Vassard, A. Krygelmans, B. Gouillieux, L. Le Gall
FRA1253	*Champia* sp3	Guadeloupe, 07/05/12	F. Rousseau, Y. Buske, J. Espinosa, L. Le Gall, E. Vassard, J. Utge, M. Caballer, R. Moolenbeek, Y. Turpin
FRA1412	*Champia* sp3	Guadeloupe, 10/05/12	L. Charles, R. Pénisson, J. Poupin, F. Rousseau, G. Dirberg, S. Soubzmaigne
FRA1830	*Champia* sp3	Guadeloupe, 18/05/12	L. Le Gall, E. Vassard
FRA1984	*Champia* sp3	Guadeloupe, 23/05/12	L. Le Gall, E. Vassard
FRA2663	*Champia* sp3	Guadeloupe, 14/12/12	F. Rousseau, Y. Buske, D. Baltide, C. Lefèbvre

Specimen identifier	Taxon	Region and date of collection	Collectors
Champiaceae			
FRA3040	*Champia* sp3	Martinique, 06/09/16	L. Le Gall, R. Ferry
FRA3047	*Champia* sp3	Martinique, 06/09/16	L. Le Gall, R. Ferry, N. Michez, M. Oliverio, M. Caballer
FRA3427	*Champia* sp3	Martinique, 14/09/16	Y. Buske, M. Oliverio, P. Stahlschmidt, L. Le Gall, R. Ferry, J. Colomb, N. de Voogd, E. Van der Ent
FRA3482	*Champia* sp3	Martinique, 14/09/16	Y. Buske, M. Oliverio, P. Stahlschmidt, L. Le Gall, R. Ferry, J. Colomb, E. Van der Ent
FRA3696	*Champia* sp3	Martinique, 17/09/16	L. Le Gall, R. Ferry, E. Van der Ent, J. Colomb
FRA1109	*Champia* sp4	Guadeloupe, 04/05/12	F. Rousseau, Y. Buske, J. Espinosa, M. Snyder, E. Vassard, J. Utge, M. Caballer, R. Moolenbeek
FRA1209	*Champia* sp5	Guadeloupe, 06/05/12	F. Rousseau, Y. Buske, J. Espinosa
LLG3298	*Champia* sp6	Pyrénées-Orientales, Languedoc-Roussillon, 05/07/10	L. Le Gall, J. Utge
LLG6069	*Champia* sp6	Corsica, 07/05/19	P. Lozouet, A.-L. Janson
LLG6121	*Champia* sp6	Corsica, 08/05/19	L. Le Gall, F. Otero Ferrer, S. Sartoretto, A. Krygelmans
LLG6516	*Champia* sp6	Corsica, 19/05/19	L. Le Gall, E. Vassard, B. Gouillieux, M. Caballer
LLG2238	*Champia* sp7	Var, Provence-Alpes-Côte d'Azur, 20/10/08	L. Le Gall, F. Rousseau
LLG5330	*Champia* sp7	Alpes-Maritimes, Provence-Alpes-Côte d'Azur, 19/06/17	F. Rousseau, L. Le Gall
FRA2184	*Champia* sp8	Guadeloupe, 28/05/12	R. Moolenbeek, L. Le Gall, E. Vassard, A. Antonioli

Specimen identifier	Taxon	Region and date of collection	Collectors
Champiaceae			
FRA1764	Champiaceae	Guadeloupe, 16/05/12	L. Le Gall, Y. Buske, Y. Turpin, M. Caballer, E. Vassard, R. Moolenbeek, K. Netchy, A. Antonioli
FRA2128	Champiaceae	Guadeloupe, 25/05/12	L. Le Gall, Y. Turpin
Lomentariacea			
FRA1945	*Ceratodictyon* sp1	Guadeloupe, 21/05/12	L. Le Gall, Y. Turpin, R. Moolenbeek, K. Netchy, E. Vassard, J. Espinosa, A. Antonioli
FRA3170	*Ceratodictyon* sp1	Martinique, 08/09/16	N. de Voogd, E. Van der Ent, N. Michez, L. Le Gall, M. Caballer
FRA3371	*Ceratodictyon* sp2	Martinique, 12/09/16	L. Le Gall, S. Schiaparelli, J. Colomb, Y. Buske, A. Krygelmans, N. Michez, M. Oliverio, R. Ferry, M. Caballer
MAD3230	*Ceratodictyon* sp2	Papua New Guinea, 19/11/12	F. Leliaert, L. Le Gall
FRA3495	*Ceratodictyon* sp3	Martinique, 14/09/16	S. Schiaparelli
MAD2311	*Ceratodictyon* sp3	Anosy, Madagascar, 10/06/10	L. Le Gall
BDR0041	*Lomentaria articulata* (Hudson) Lyngbye	Finistère, Brittany, 23/07/09	B. de Reviers
BDR0042	*Lomentaria articulata*	Finistère, Brittany, 23/07/09	B. de Reviers
LLG1314	*Lomentaria articulata*	Channel, Normandy, 27/05/08	L. Le Gall, S. Petinay
LLG1611	*Lomentaria articulata*	Ille-et-Vilaine, Brittany, 15/07/08	L. Le Gall, J. Utge, F. Rousseau
LLG1613	*Lomentaria articulata*	Ille-et-Vilaine, Brittany, 15/07/08	L. Le Gall, J. Utge, F. Rousseau
LLG1830	*Lomentaria articulata*	Ille-et-Vilaine, Brittany, 17/07/08	L. Le Gall, F. Rousseau, J. Utge

Specimen identifier	Taxon	Region and date of collection	Collectors
Lomentariacea			
LLG2933	*Lomentaria articulata*	Finistère, Brittany, 11/08/09	L. Le Gall, J. Utge, Y. Turpin
LLG2934	*Lomentaria articulata*	Finistère, Brittany, 11/08/09	L. Le Gall, J. Utge, Y. Turpin
LLG3274	*Lomentaria articulata*	Finistère, Brittany, 20/05/10	L. Le Gall, Y.N. Fontana
LLG4453	*Lomentaria articulata*	Côtes-d'Armor, Brittany, 10/07/12	L. Le Gall
RMAR1063	*Lomentaria articulata*	Finistère, Brittany, 15/03/11	L. Le Gall, S. Brégéon
LLG3889	*Lomentaria* cf *firma* (J.Agardh) Falkenberg	Sicily, 18/04/11	J. Utge, L. Le Gall
LLG2450	*Lomentaria clavellosa* (Turner) Gaillon	Channel, Normandy, 14/03/09	L. Le Gall
LLG3262	*Lomentaria clavellosa*	Finistère, Brittany, 20/05/10	L. Le Gall
LLG4080	*Lomentaria clavellosa*	Finistère, Brittany, 30/05/11	L. Le Gall
LLG5121	*Lomentaria clavellosa*	Finistère, Brittany, 01/07/13	L. Le Gall
RMAR1062	*Lomentaria clavellosa*	Finistère, Brittany, 15/03/11	L. Le Gall, S. Brégéon
TJS0012	*Lomentaria clavellosa*	Finistère, Brittany, 11/07/07	T. Silberfeld
LLG0595	*Lomentaria hakodatensis* Yendo	Côtes-d'Armor, Brittany, 28/09/07	L. Le Gall, J. Utge, L. Bittner, T. Silberfeld
PML0141	*Lomentaria orcadensis* (Harvey) Collins	Finistère, Brittany, 20/09/14	L. Le Gall
PML0155	*Lomentaria orcadensis*	Finistère, Brittany, 20/09/14	L. Le Gall

Specimen identifier	Taxon	Region and date of collection	Collectors
		Lomentariacea	
PML0156	*Lomentaria orcadensis*	Finistère, Brittany, 20/09/14	L. Le Gall
PML0157	*Lomentaria orcadensis*	Finistère, Brittany, 20/09/14	L. Le Gall
PML0223B	*Lomentaria orcadensis*	Finistère, Brittany, 22/06/15	L. Le Gall, L. Lévêque
PML0241B	*Lomentaria orcadensis*	Finistère, Brittany, 23/06/15	L. Le Gall, L. Lévêque
PML0250	*Lomentaria orcadensis*	Finistère, Brittany, 23/06/15	L. Le Gall, L. Lévêque
PML0295	*Lomentaria orcadensis*	Finistère, Brittany, 11/09/15	L. Le Gall
LLG4635	*Lomentaria* sp1	Andalusia, 21/09/12	L. Le Gall
LLG6634	*Lomentaria* sp1	Corsica, 22/05/19	E. Vassard, A. Krygelmans, B. Gouillieux, L. Le Gall
FRA1978	*Lomentaria* sp2	Guadeloupe, 23/05/12	L. Le Gall, E. Vassard
FRA3699	*Lomentaria* sp2	Martinique, 17/09/16	L. Le Gall, R. Ferry, E. Van der Ent, J. Colomb
FRA1962	*Lomentaria* sp3	Guadeloupe, 23/05/12	L. Le Gall, E. Vassard
MAD0557	*Lomentaria* sp4	Anosy, Madagascar, 11/05/10	F. Rousseau
MAD3896	*Lomentaria* sp5	Madang, Papua New Guinea, 06/12/12	F. Leliaert, L. Le Gall
LLG2162	*Lomentaria subdichotoma* Ercegovic	Var, Provence-Alpes-Côte d'Azur, 19/10/08	L. Le Gall, Y. Turpin
LLG2178	*Lomentaria subdichotoma*	Var, Provence-Alpes-Côte d'Azur, 19/10/08	L. Le Gall, Y. Turpin

Table 8.1. *List and collection information of the specimens included in our study. The identifiers of the specimens present in the phylogenetic trees are highlighted in bold*

8.2.2. *Acquisition of molecular data*

DNA extraction is performed for each specimen using commercial kits (Vergés *et al.* 2014b), following the manufacturer's recommendations. The gene fragments of interest are amplified by PCR and sequenced following the Sanger method. The mitochondrial gene coding for cytochrome oxidase 1 (COI) was selected as the initial marker for the primary hypothesis of species delineation, on the basis of DNA sequences.

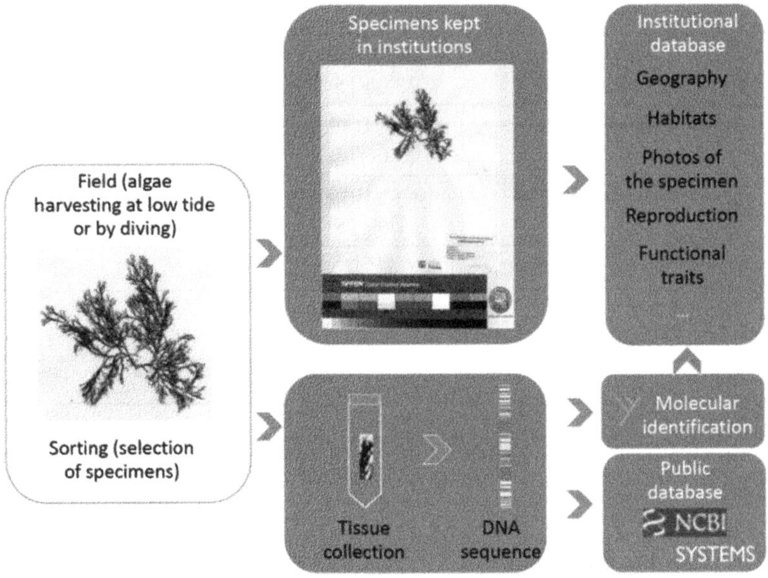

Figure 8.3. *Chain of analysis of specimens from the field up until the acquisition of DNA sequences. For a color version of this figure, see www.iste.co.uk/grandcolas/systematics.zip*

8.2.3. *Analysis of phylogenetic relationships between Atlantic and Mediterranean specimens*

We searched among our nearly 4,000 COI sequences of red algae (including 416 specimens from the Mediterranean and 287 from the Atlantic), with the recently diverging groups corresponding more or less at the genus level, and presenting specimens from the Atlantic and Mediterranean. Among these groups, we chose to focus on two genera of Rhodymeniales, *Champia* and *Lomentaria*, which appeared to present different relational trajectories between the Atlantic and Mediterranean.

We extracted 41 sequences for the genus *Lomentaria* and 31 for *Champia* from our molecular database (Table 8.1, section 8.2.1). Among identical sequences, we selected one sequence using the "Perl collapsetypes_v4.6.pl" script. Ultimately, the two alignment sets were composed of 21 sequences and 588 sites for *Lomentaria* and 22 sequences and 599 sites for *Champia*.

For phylogenetic reconstructions and a rough estimation of node divergence times, we used the same parameters as those used in the Bringloe and Saunders (2018) study. These authors inferred calibrated phylogenies on some red algae genera, using different molecular markers including CO1, in order to discuss the origin of speciation events after the opening of the Bering Strait.

Thus, for each of the two datasets, a molecular phylogeny was reconstructed using the maximum likelihood method by using the *PhyML* software (Guindon *et al.* 2010), implemented in SeaView (Gouy *et al.* 2010). The HKY85 molecular evolution model was selected, and 1,000 bootstrap runs were performed to test the sturdiness of the tree. The estimation of node divergence times was performed using the *BEAST 1.10.4* software (Drummond *et al.* 2012; Suchard *et al.* 2018) by imposing the topology obtained in ML. The same parameters as those of Bringloe and Saunders (2018) were applied, including constant molecular clock and mutation rate estimation of 0.0076 substitution/site/Ma following a normal distribution with a standard deviation of 0.005.

8.3. Results

For each of the two taxa studied, a calibrated tree was obtained. Figure 8.4 for *Champia* and Figure 8.5 for *Lomentaria* represent these trees with the estimated mean time of divergence at the nodes and the uncertainties in the form of 95% credibility intervals (HPD). The main objective of the calibration here is to have an estimate of the age of diversification of Mediterranean species and to verify that it took place after the Zanklean transgression. Since we could not rely on paleontological data, we used a strict molecular clock and parameters from recent work on red algae (Bringloe and Saunders 2018). The values of the estimated ages must therefore be considered as rough estimates and will have to be refined later. We also note that the credibility intervals are broad around the estimated ages.

For *Champia*, around the European coasts of the Celtic Sea and the Mediterranean, the analyses highlight the presence of at least four entities that present levels of divergence of the order of those observed between species. Among these four species, only one is present in the Atlantic and three are in the Mediterranean. This group of four species of *Champia*, which is strongly supported, is a sister group to a species from the Caribbean. The biogeographic trajectory of the European *Champia* thus appears to originate from a *Champia* of the tropical Atlantic region. The resulting diversification into three Mediterranean species, and one in the Celtic Sea, took place after the reopening of the Strait of Gibraltar. It is interesting to note that within this group of four species, the first divergences are represented by the Mediterranean species and that the Atlantic *Champia* sp1 has the Mediterranean *Champia* sp2 as its sister group.

For the *Lomentaria* dataset, the analyses highlight the presence of at least seven entities with levels of divergence of the order of those observed between species. Although the specimen of *Lomentaria hakodatensis* included in our dataset comes from the Atlantic, this species is nevertheless considered to be Pacific. Indeed, it probably would have been introduced into Europe, in the Mediterranean (Verlaque *et al.* 2015) and the Atlantic (Loiseaux-de Goër and Noailles 2008), in connection with Japanese oyster farming on the European coasts. The three species recorded so far on the northeast Atlantic coasts are included in our dataset: *Lomentaria articulata*, *L. orcadensis* and *L. clavellosa*. All three are congruent with a lineage that is strongly supported by phylogenetic analyses and which also contains two other Mediterranean species, *Lomentaria* sp1 and *L. subdichotoma*. These two Mediterranean species are thought to share a common ancestor with *L. clavellosa*, whose node age is estimated at 5.64 million years. Contrary to *Champia*, within this group of North Atlantic and Mediterranean species, the first divergences are Atlantic species. The trajectory of this lineage thus appears to have a temperate Atlantic origin, with Mediterranean diversification after the re-opening of Gibraltar. Despite the fact that *L. articulata* and *L. clavellosa* have been recorded in the Mediterranean, we have not identified these two species among the specimens in our dataset. We have highlighted another lineage containing Caribbean Atlantic species and a Mediterranean *Lomentaria* species, *Lomentaria* cf *firma*, whose node age in common with *Lomentaria* sp2 is estimated at 5.07 Ma.

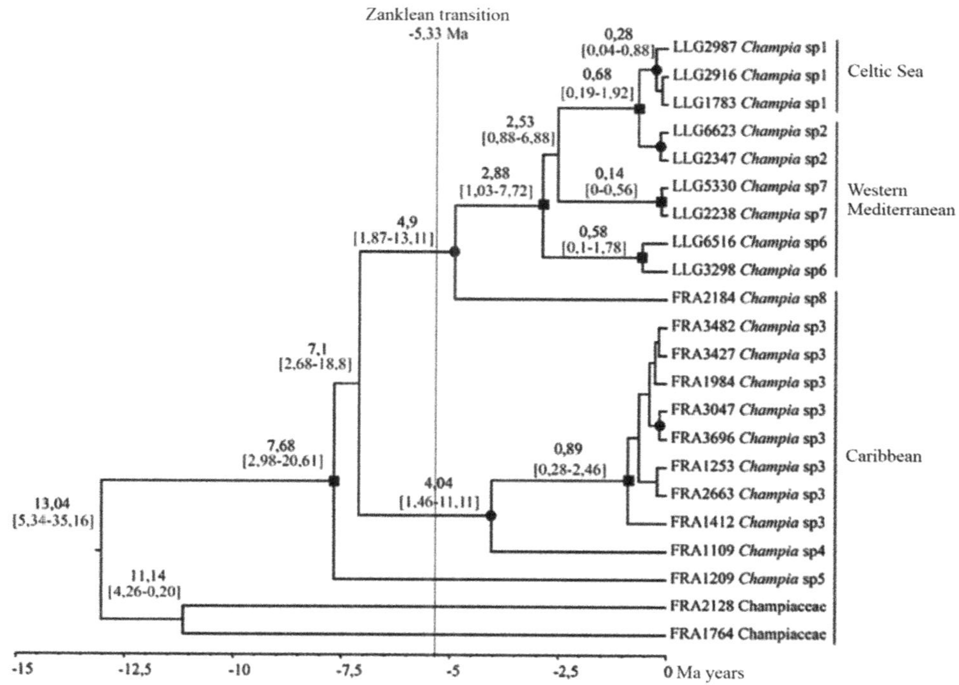

Figure 8.4. *Calibrated* Champia *tree obtained with BEAST. For nodes above the specific level, the estimated value of the mean divergence time is presented with uncertainties between two brackets as 95% credibility intervals (HPD). The bootstrap values obtained with PhyML are represented either by a black square at nodes with values greater than or equal to 95%, or by a black circle for values between 70% and 95%*

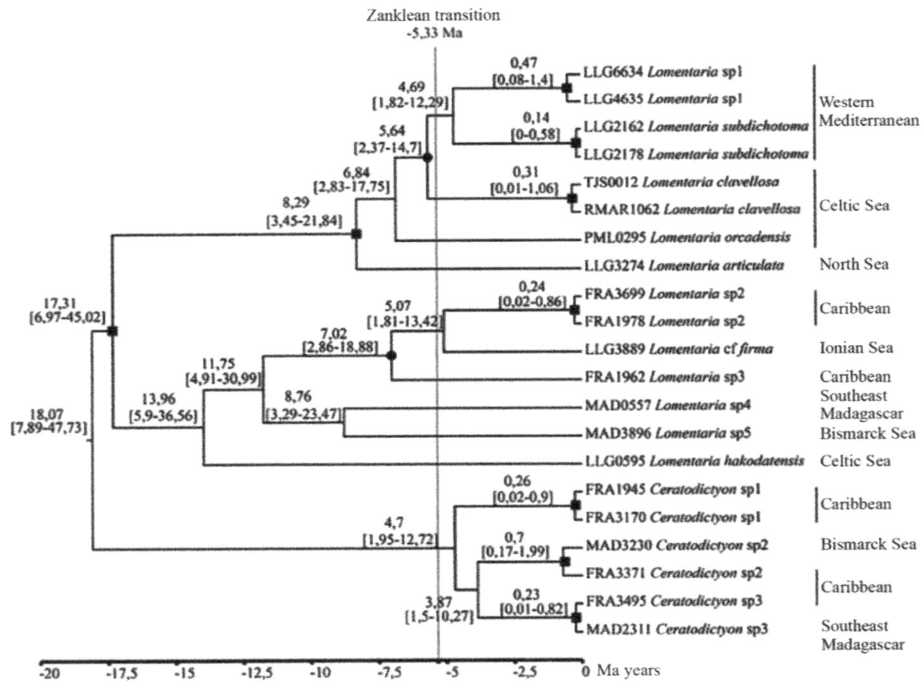

Figure 8.5. *Calibrated* Lomentaria *tree obtained with BEAST. For nodes above the specific level, the estimated value of the mean divergence time is presented with uncertainties between two brackets as 95% credibility intervals (HPD). The bootstrap values obtained with PhyML are represented either by a black square for nodes with values greater than or equal to 95%, or by a black circle for values between 70% and 95%*

8.4. Discussion

The Mediterranean transition at the biological level is controversial, as there are conflicting results regarding the biogeographical separation between Atlantic and Mediterranean biota. The opening of the Strait of Gibraltar, which put an end to the Messinian salinity crisis, allowed a connection to be established between the marine organisms of the Atlantic Ocean and the Mediterranean Sea. At present, the evaporation of Mediterranean waters leads to a deficit in the water balance of this sea. However, exchanges with the Atlantic at the level of the Strait of Gibraltar allow this balance to be evened and result in the inflow of oceanic waters on the surface, and the discharge of more salty Mediterranean waters into the Atlantic at depth.

On the biogeographic level, after the reopening of the Strait of Gibraltar, several geological events influenced the history of the Northeast Atlantic region (in the sense of Spalding (2007), which includes the Mediterranean) such as the most recurrent glaciations during the Pleistocene. Analysis of our datasets tends to show a clear separation between Atlantic and Mediterranean species, as we have not found any species present in both regions. Our two examples confirm Feldmann's hypothesis that the Atlantic element (both in its boreal and tropical entity) is the main element at the origin of the Mediterranean algal flora. Nevertheless, the results obtained on *Lomentaria* and *Champia* suggest that there is not a single pattern of relationships between Mediterranean and Atlantic red algae, but that there is likely a mosaic of trajectories, which is consistent with the analyses carried out by Patarnello (2007) on various marine organisms.

The methodology proposed here, based on dense sampling and identifications carried out using tools of molecular systematics, has proven to be suitable for revealing relationships and discussing trajectories. In order to refine the relative importance of each element and thus better understand the origin of the diversity of Mediterranean algae, it would be beneficial to extend this study to all algae. The dataset available will allow such a study to be carried out; however, it would be desirable to intensify sampling along the African coasts of the Atlantic and Mediterranean.

8.5. Acknowledgments

We thank ISYEB for the invitation to present the results of this study during the *Journée scientifique de l'ISYEB* (ISYEB Science Day) held at the museum on October 3, 2019. Our sincere thanks are addressed to all the collectors of the studied specimens. Most of the specimens were collected during expeditions, notably Atimo Vatae, Karubenthos, Papua, Madibenthos, Corsica, Brittany, Normandy (SMEL) and Villefranche-sur-Mer.

The Atimo Vatae expedition to southern Madagascar (coordinator: Philippe Bouchet) is part of a group of Mozambique–Madagascar expeditions financed by the Total Foundation, the Prince Albert II of Monaco Foundation and the Stavros Niarchos Foundation under the initiative *La planète revisitée* (The Planet, Revisited in English), a joint initiative of the French National Museum of Natural History (MNHN) and Pro Natura International (PNI), in partnership with the Institute of Fisheries and Marine Sciences of the University of Toliara (IH.SM) and the Madagascar office of the Wildlife Conservation Society (WCS). The Institute of Research for Development (IRD) deployed Antea, its research catamaran.

The Karubenthos Expedition (coordinator: Philippe Bouchet) was jointly organized by the Guadeloupe National Park, the MNHN, the University of the West Indies and Guyana (UAG) and the Pierre and Marie Curie University (UPMC), with funding from the European Regional Development Fund (ERDF) and the Autonomous Port of Guadeloupe.

The specimens of the Madang expedition were obtained during the *La planète revisitée* expedition in Papua New Guinea (principal investigators Philippe Bouchet, Claude Payri and Sarah Samadi), organized by the MNHN, PNI, the IRD and the University of Papua New Guinea (UPNG). The organizers would like to thank the Total Foundation, the Prince Albert II of Monaco Foundation, the EDF Foundation, the Stavros Niarchos Foundation and Entrepose Contracting for their funding, as well as the Divine Word University (DWU) for its in-kind support. The expedition took place under a permit issued by the Department of Environment and Conservation of Papua New Guinea.

The Madibenthos expedition (principal researcher: Philippe Bouchet) was organized by the MNHN with the support of the ERDF, the Agency for Marine Protected Areas (AAMP), the DEAL and the Water Office (ODE) of Martinique, as well as the logistical support of the French Navy via the reception at the Fort Saint-Louis base. The specimens collected at Villefranche-sur-Mer were supported by the EMBRC.

The *La planète revisitée* in Corsica expedition was organized with the support of the natural marine park of Cap Corse and Agriate, the University of Corsica Pasquale Paoli and the Corsican Environment Office (OEC), with funding from the Corsican local authority and the French Office for Biodiversity (OFB). The sequencing was carried out at the molecular systematics department of the MNHN (CNRS – UMS 2700), with funds provided by the museum's transversal action. Gilberto Marani ensures the digitization of the specimens and the updating of the database, which we thank him for here.

8.6. References

Adl, S.M., Bass, D., Lane, C.E., *et al.* (2019). Revisions to the classification, nomenclature, and diversity of eukaryotes. *Journal of Eukaryotic Microbiology*, 66, 4–119.

Benhissoune, S., Boudouresque, C.-F., and Verlaque, M. (2001). A check-list of marine seaweeds of the Mediterranean and Atlantic coasts of Morocco I. Chlorophyceae Wille s. l. *Botanica Marina*, 44(2) [Online]. Available: https://www.degruyter.com/view/j/botm.2001.44.issue-2/bot.2001.023/bot.2001.023.xml [Accessed 20 December 2019].

Benhissoune, S., Boudouresque, C.-F., and Verlaque, M. (2002). A checklist of the seaweeds of the Mediterranean and Atlantic coasts of Morocco II. Phaeophyceae. *Botanica Marina*, 45(3), 217–230 [Online]. Available: https://www.degruyter.com/view/journals/botm/45/3/article-p217.xml.

Boudouresque, C.F. (2004). Marine biodiversity in the mediterranean: Status of species, populations and communities. *Scientific Reports of Port-Cros National Park*, 20, 97–146.

Bringloe, T. and Saunders, G. (2018). Mitochondrial DNA sequence data reveal the origins of postglacial marine macroalgal flora in the Northwest Atlantic. *Marine Ecology Progress Series*, 589, 45–58.

Bunker, F.S.D., Brodie, J.A., Maggs, C., and Bunker, A.R. (2017). *Seaweeds of Britain and Ireland*, 2nd edition. Wild Nature Press, Plymouth.

Coll, M., Piroddi, C., Steenbeek, J. *et al.* (2010). The biodiversity of the Mediterranean Sea: Estimates, patterns, and threats. *PLOS One*, 5(8), e11842.

De Jode, A., David, R., Haguenauer, A., Cahill, A.E., Erga, Z., Guillemain, D., Sartoretto, S., Rocher, C., Selva, M., Le Gall, L., Féral, J.-P., Chenuil, A. (2019). From seascape ecology to population genomics and back. Spatial and ecological differentiation among cryptic species of the red algae *Lithophyllum stictiforme/L. cabiochiae*, main bioconstructors of coralligenous habitats. *Molecular Phylogenetics and Evolution*, 137, 104–113.

Fredj, G., Bellan Santini, D., and Meinardi, M. (1992). État des connaissances sur la faune marine méditerranéenne. *Bulletin de l'Institut océanographique de Monaco*, 9, 133–145.

Giaconne, G. (1974). Tipologia delle comunita' fitobentoniche del Mediterraneo. *Memorie di biologia marina e di oceanografia*, 4(4–6), 149–168.

Giaconne, G. and Geraci, R.M. (1989). Biogeografia delle alghe del Mediterraneo. *Anales del Jardín Botánico de Madrid*, 46(1), 27–34.

Gómez Garreta, A., Gallardo, T., Ribera, M.A., Cormaci, M., Furnari, G., Giaccone, G., and Boudouresque, C.F. (2001). Checklist of Mediterranean seaweeds III. Rhodophyceae Rabenh 1. Ceramiales Oltm. *Botanica Marina*, 44(5) [Online]. Available: http://www.degruyter.com/view/j/botm.2001.44.issue-5/bot.2001.051/bot.2001.051.xml [Accessed 23 January 2016].

Gouy, M., Guindon, S., and Gascuel, O. (2010). SeaView Version 4: A multiplatform graphical user interface for sequence alignment and phylogenetic tree building. *Molecular Biology and Evolution*, 27(2), 221–224.

Guindon, S., Dufayard, J.-F., Lefort, V., Anisimova, M., Hordijk, W., and Gascuel, O. (2010). New algorithms and methods to estimate maximum-likelihood phylogenies: Assessing the performance of PhyML 3.0. *Systematic Biology*, 59(3), 307–321.

Guiry, M.D. and Guiry, G.M. (2020). AlgaeBase. Database, National University of Ireland, Galway [Online]. Available: http://www.algaebase.org [Accessed 22 June 2020].

Guiry, M.D., Guiry, G.M., Morrison, L., Rindi, F., Miranda, S.V., Mathieson, A.C., Parker, B.C., Langangen, A., John, D.M., Bárbara, I., Carter, C.F., Kuipers, P., and Garbary, D.J. (2014). AlgaeBase: An online resource for algae. *Cryptogamie, Algologie*, 35(2), 105–115.

Hebert, P.D.N., Cywinska, A., Ball, S.L., and deWaard, J.R. (2003). Biological identifications through DNA barcodes. *Proceedings of the Royal Society B: Biological Sciences*, 270(1512), 313–321.

Keith, S.A., Kerswell, A.P., and Connolly, S.R. (2014). Global diversity of marine macroalgae: Environmental conditions explain less variation in the tropics: Global diversity of marine macroalgae. *Global Ecology and Biogeography*, 23(5), 517–529.

Kerswell, A.P. (2006). Global biodiversity patterns of benthic marine algae. *Ecology*, 87(10), 2479–2488.

Le Gall, L., Delsuc, F., Hourdez, S., Lecointre, G., and Rasplus, J.-Y. (2017). Toward the DNA Library of Life. *European Journal of Taxonomy*, 266(2017) [Online]. Available: http://www.europeanjournaloftaxonomy.eu/index.php/ejt/article/view/390 [Accessed 30 January 2017].

Le Gall, L., Gey, D., and Vergés, A. (2018). Insights into the rare Mediterranean endemic *Kallymenia spathulata*: DNA phylogenies resolve this species as Halymeniaceae (Halymeniales) rather than Kallymeniaceae (Gigartinales), with the proposal of *Felicinia spathulata* comb. nov. *Cryptogamie, Algologie*, 39(3), 339–347.

Leliaert, F., Payo, D.A., Gurgel, C.F.D., Schils, T., Draisma, S.G.A., Saunders, G.W., Kamiya, M., Sherwood, A.R., Lin, S.-M., Huisman, J.M., Le Gall, L., Anderson, R.J., Bolton, J.J., Mattio, L., Zubia, M., Spokes, T., Vieira, C., Payri, C.E., Coppejane, E., D'hondt, S., Verbruggen, H., De Clerck, O. (2018). Patterns and drivers of species diversity in the Indo-Pacific red seaweed *Portieria*. *Journal of Biogeography* [Online]. Available: https://onlinelibrary.wiley.com/doi/abs/10.1111/jbi.13410 [Accessed 9 August 2018].

Loiseaux-de Goër, S. and Noailles, M.-C. (2008). *Algues de Roscoff*. Éditions de la Station biologique de Roscoff, Roscoff.

Manghisi, A., Le Gall, L., Ribera, M.A., Bonillo, C., Gargiulo, G.M., and Morabito, M. (2014). The Mediterranean endemic new genus *Felicinia* (Halymeniales, Rhodophyta) recognized by a morphological and phylogenetic integrative approach. *Cryptogamie, Algologie*, 35(3), 221–243.

Mora, C., Tittensor, D.P., Adl, S., Simpson, A.G.B., and Worm, B. (2011). How many species are there on Earth and in the ocean? *PLOS Biology*, 9(8), e1001127.

Patarnello, T., Volckaert, F.A.M.J., and Castilho, R. (2007). Pillars of Hercules: Is the Atlantic-Mediterranean transition a phylogeographical break? Phylogeography of the mediterranean-atlantic divide. *Molecular Ecology*, 16(21), 4426–4444.

Peña, V., De Clerck, O., Afonso-Carrillo, J., Ballesteros, E., Bárbara, I., Barreiro, R., and Le Gall, L. (2015). An integrative systematic approach to species diversity and distribution in the genus *Mesophyllum* (Corallinales, Rhodophyta) in Atlantic and Mediterranean Europe. *European Journal of Phycology*, 50(1), 20–36.

Ponce-Toledo, R.I., Deschamps, P., López-García, P., Zivanovic, Y., Benzerara, K., and Moreira, D. (2017). An early-branching freshwater cyanobacterium at the origin of plastids. *Current Biology*, 27(3), 386–391.

Rindi, F., Braga, J.C., Martin, S., Peña, V., Le Gall, L., Caragnano, A., and Aguirre, J. (2019). Coralline algae in a changing Mediterranean Sea: How can we predict their future, if we do not know their present? *Frontiers in Marine Science*, 6, 723.

Rodríguez-Prieto, C. (2013). *Guía de las macroalgas y fanerógamas marinas del Mediterráneo Occidental*. Omega, Barcelona.

Saunders, G.W. (2005). Applying DNA barcoding to red macroalgae: A preliminary appraisal holds promise for future applications. *Philosophical Transactions of the Royal Society B: Biological Sciences*, 360(1462), 1879–1888.

Schultz, N.E., Lane, C.E., Le Gall, L., Gey, D., Bigney, A.R., De Reviers, B., Rousseau, F., and Schneider, C.W. (2015). A barcode analysis of the genus *Lobophora* (Dictyotales, Phaeophyceae) in the western Atlantic Ocean with four novel species and the epitypification of *L. variegata* (J.V. Lamouroux) E.C. Oliveira. *European Journal of Phycology*, 50(4), 481–500.

Stambler, N. (ed.) (2012). *Life in the Mediterranean Sea: A Look at Habitat Changes*. Nova Science, New York.

Suchard, M.A., Lemey, P., Baele, G., Ayres, D.L., Drummond, A.J., and Rambaut, A. (2018). Bayesian phylogenetic and phylodynamic data integration using BEAST 1.10. *Virus Evolution*, 4(1) [Online]. Available: https://academic.oup.com/ve/article/doi/10.1093/ve/vey016/5035211 [Accessed 26 December 2019].

Tittensor, D.P., Mora, C., Jetz, W., Lotze, H.K., Ricard, D., Berghe, E.V., and Worm, B. (2010). Global patterns and predictors of marine biodiversity across taxa. *Nature*, 466(7310), 1098–1101.

Vergés, A., Gey, D., Utgé, J., Cruaud, C., and Le Gall, L. (2014a). Recognition of a new species of *Kallymenia* (Gigartinales, Rhodophyta) from Croatia (Mediterranean Sea) based on morphology and DNA barcode. *European Journal of Phycology*, 49(3), 332–344.

Vergés, A., Gey, D., Utgé, J., Cruaud, C., and Le Gall, L. (2014b). Recognition of a new species of *Kallymenia* (Gigartinales, Rhodophyta) from Croatia (Mediterranean Sea) based on morphology and DNA barcode. *European Journal of Phycology*, 49(3), 332–344.

Verlaque, M., Ruitton, S., Mineur, F., and Boudouresque, C.F. (2015). *CIESM Atlas of Exotic Species of the Mediterranean 4. Macrophytes*. CIESM Publishers, Monaco.

Vieira, C., Camacho, O., Wynne, M.J., Mattio, L., Anderson, R.J., Bolton, J.J., Sansón, M., D'hondt, S., Leliaert, F., Fredericq, S., Payri, C., and De Clerck, O. (2016). Shedding new light on old algae: Matching names and sequences in the brown algal genus *Lobophora* (Dictyotales, Phaeophyceae). *Taxon*, 65(4), 689–707.

9

Ontogeny and Evolution of the Hyperorgan of Delphinieae

A hyperorgan is the morphologically observable result of the synorganization of parts of an organism into a functional unit. In this chapter, we define the concept of synorganization and give the history of its use, before developing the case of the floral hyperorgan of the Delphinieae (Ranunculaceae). We present state of the art research on the disparity, ontogeny and evolution of this hyperorgan within the tribe. Finally, we place the concept of synorganization in the general framework of evolutionary biology studies.

9.1. Introduction

Work on floral synorganization in the tribe of Delphinieae (Ranunculaceae) started in the "Plant Morphology, Anatomy and Development" team at the National Museum of Natural History, Paris (MNHN) in 2018; since then, several students have addressed different aspects of this topic. Their names appear in the list of co-authors in this chapter, in chronological order of their visits to the laboratory. They have all contributed to data acquisition and our understanding of the concept discussed here.

By its etymology, the term *synorganization* (from the Greek σύν, together, and ὄργανον, body organ, musical instrument, voice) could *a priori*

Chapter written by Florian JABBOUR, Julie ZALKO, Antoine MOREL, Samuel FRACHON and Isabelle BOUCHART-DUFAY.

designate a group of organs acting together, or, if one prefers a musical metaphor, an orchestra playing the same music, or a group of voices singing in a choir. We can already see that this concept is rich and covers several aspects: spatial, temporal (short and long time) and functional. The basic elements fit together spatially during ontogeny, and form a functional structure subject to natural selection. Starting from this intuitive definition, we will return to the origin of the concept and trace the evolution of its contents and contours from the middle of the 20th century. This theoretical section will be followed by the presentation of a case study in evolutionary botany. In the final section, studies of plant morphology on the theme of synorganization will be placed in a context of evolutionary biology.

9.2. Synorganization: a concept, definitions

9.2.1. *Adolf Remane and the synorganization of animal structures*

It was in 1952, in his treatise entitled Die Grundlagen des Natürlichen Systems, der Vergleichenden Anatomie und der Phylogenetik (Bases of Natural Systems, Comparative Anatomy and Phylogenetics), that Adolf Remane introduced the term synorganization (Remane 1956). As professor and director of the Zoological Institute and Museum in Kiel (Germany), he was interested in paleontology, marine ecology and animal morphology (Zachos and Hoßfeld 2006). Here is his definition of this term, translated into English:

> Association of different parts of an organism, without any particular link between them, into a complex new organ, a functional system or "apparatus", and the thorough and harmonious development of this apparatus. [...] The function of a synorganized structure is physiological–biological.

He exclusively relies on animal examples to illustrate his point, such as the reflex arc (composed of sensory, signal processing and motor neurons) or the *Drepanosiphum platanoidis* aphid's front and back wing adhesion device.

9.2.2. *A concept adopted by botanists, and by flower specialists in particular*

The concept of synorganization does not seem to have spread in the field of zoology (see, however, Salvini-Plawen and Splechtna (1979) and Wagner (1989)). However, it was taken up by botanists at a very early stage (Vogel 1959, 1969), and the definition of the term has been further developed and refined in the course of studies on the various taxa of angiosperms (Endress 1990; Specht and Bartlett 2009; Endress 2010; Ronse De Craene 2010; Ley and Claßen-Bockhoff 2012; Endress 2016). A consensual definition would be: *spatial association, integration, interaction, connection of individual parts of an organism or ancestrally independent, neighboring, similar or different types of organs, with or without fusion, in a complex structure, a functional unit, a hyperorgan.*

Synorganization studies have focused on floral structures. The hyperorgan is usually visible to the naked eye, for example, the orchid gynostem, which is a column resulting from the fusion of the male and female sexual organs, which allows a precise deposition of pollen on both the body of the pollinator and the stigma. The nectariferous pseudotubes of *Geranium robertianum* are another example. These are spaces in which the pollinator sucking appendage is inserted; they are produced by the juxtaposition of the outer walls of six related organs (Endress 2010).

9.2.3. *A concept to be limited organically, and to be placed in a phylogenetic framework*

You might tell us that organizations are full of functional systems that come from the combination of various elements. Indeed, whether we consider the double helix of DNA or an organism as a whole, everything could be interpreted as a synorganization. If we want to manipulate this concept, it is necessary to limit it at the organic level. Between the extremes of the atom and the whole organism (note that in clonal or symbiotic plants, defining an organism is a challenge), we propose adding that synorganization implies macroscopic structures, organs or fragments of organs, to the definition stated above. Moreover, *the constituent structures of a hyperorgan can only be qualified as synorganized in comparison with homologous non-synorganized structures in a phylogenetically close taxon.*

9.3. Ontogeny and evolution of the hyperorgan of Delphinieae

9.3.1. *Disparity of the hyperorgan in the tribe*

Within the Ranunculaceae family (order Ranunculales, sister group of all other eudicots (APG 2016)), the Delphinieae lineage is the only one where zygomorphy has evolved. The bilateral symmetry of the Delphinieae flower is the result of the differential development of the organs of the dorsal and ventral halves (Figure 9.1a). The two most dorsal petals are spurred, and produce and store nectar. While it is not uncommon to find pairs of spurs (nectariferous as in *Satyrium*, or oil-bearing as in *Diascia*) within angiosperms, a pair of spurs nested in a third is much rarer. Only the 600 or so species of Delphinieae are endowed with them, and this morphological signature must have already been present in the common ancestor of all the species of the tribe.

The spur of the dorsal sepal envelops the nectariferous spurs of the two most dorsal petals (Figure 9.1b). It protects them (not always effectively) from attack by nectar-stealing insects (Figure 9.1c). It may also create conditions of temperature and humidity, a sort of microclimate, allowing the nectar produced and stored in the inner spurs to be preserved.

Figure 9.1. *Flowers of* Delphinium *L. (Ranunculaceae). For a color version of this figure, see www.iste.co.uk/grandcolas/systematics.zip*

COMMENT ON FIGURE 9.1. – *a) Flowers of D. dasycaulon Fresen. The frontal view shows the calyx composed of petaloid sepals surrounding the corolla, reduced to four dorsal petals. b) Flower of D. elatum L., from which the spurred dorsal sepal has been isolated to show the two most dorsal petals, which have been spurred themselves. c) Side view of a flower of D. dasycaulon, whose dorsal sepal spur has been pierced by an insect to steal the nectar produced and stored in the pair of internal spurs (photos by Florian Jabbour, taken in Malawi (a and c) and at the Jardin des Plantes, Paris (b)).*

The hyperorgan of the Delphinieae is thus composed of three organs, whose outermost spur envelops those of the two innermost ones. This hyperorgan plays a role in pollination, only allowing a certain range of pollinators access to nectar; those whose organ used to collect nectar is similar in size to the internal spurs (Figure 9.2). However, the connection between the basic elements of the hyperorgan does not stop there. At the level of the insertion of the dorsal petals into the floral receptacle, a more or less important fusion of the contiguous edges of the base of the two petals can be observed. Finally, in the spurs of the two dorsal petals, from the area of insertion into the receptacle and up to a certain distance, the facing walls of both spurs are indented, and the two indentations facing each other create a cavity that is common to both spurs (Figure 9.2). Floral synorganization in the Delphinieae involves three aspects: interlocking, fusion and creation of a common space.

This organization plan of the floral hyperorgan of Delphinieae comes in multiple aspects (spurs, fusion zone, common cavity), shapes and colors, and explains the great majority of the floral disparity in the tribe.

It should be noted that the length needed for a pollinator to access the nectar is neither the length of the external spur (in other words, the spur of the dorsal sepal; length traditionally measured, because it is directly accessible, and reported in taxonomic treatments), nor that of the internal spurs (in other words, the spur of the most dorsal petals), but the latter from which all or part of the length of the common cavity has been subtracted (Figure 9.2).

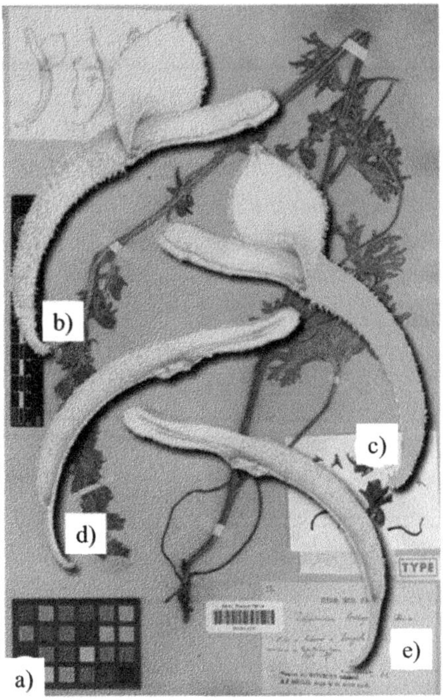

Figure 9.2. *The pair of indented spurs included in a third spur (*Delphinium *flower). For a color version of this figure, see www.iste.co.uk/grandcolas/systematics.zip*

COMMENT ON FIGURE 9.2. – *a) Specimen of J.A. Soulié 23 (P00201173), type of D. trichophorum Franch. including illustration*[1]*. b) Hyperorgan (drawing isolated from illustration) composed of the dorsal sepal whose spur – colored green in (c) – encompasses the spurs of the two dorsal petals. d) Dorsal petal with indented spur – the non-indented portion is colored green in (e).*

As morphologists and systematists, we are aware of the interest that research on the morphological diversity of the hyperorgan will have on the improvement of species descriptions, and on the re-assessment of the close relationships between the Delphinieae species and associated pollinator species.

1 The high-resolution image of the specimen is available at: http://coldb.mnhn.fr/catalognumber/mnhn/p/p00201173.

9.3.2. *Ontogeny of the synorganized structure*

The implementation of the elements involved in the hyperorgan during development is rich in lessons about the morphological diversity observed in the adult stage. Without developmental studies, it would not have been possible to confirm that the dorsal petal of *Consolida* resulted, at a very early stage of development, from the postgenital fusion of the primordia of the two most dorsal petals (Jabbour and Renner 2012a). Partial fusions of the base of the dorsal petals are also easily observable and quantifiable when scanning electron microscopy images. Following the sequence of development of Delphinieae floral buds is the way to decompose the cellular proliferation and growth processes involved in the elongation of the internal and external spurs. It appears that the formation of the external spur of the hyperorgan is preceded by that of the internal spurs. The genetic and hormonal bases of the formation of the interlocking spurs are not yet clarified, let alone the physical constraints that are potentially exerted by the internal spurs on the external spur.

9.3.3. *Evolving trends and convergences*

By combining our knowledge of the morphological diversity of the Delphinieae and the phylogeny of the tribe, we are able to reconstruct the sketch of the common ancestor of the Delphinieae, albeit incompletely. This one must have had zygomorphic flowers and a dorsoventral corolla whose two most dorsal petals were spurred and nectariferous, the spurs being covered by the spur of the dorsal sepal. During the evolution of the Delphinieae, the dorsal sepal evolved into various forms, ranging from a spurred structure, to a helmet, to a narrow cap (Figure 9.3) (Jabbour and Renner 2012a). The helmet- or cap-shaped sepals of flowers of the genus *Aconitum* could act as the common hole for the two dorsal petals of *Delphinium* and *Staphisagria*. In this first genus, the spurs of the petals are placed high, at the top of a long pedicel.

Cap-shaped sepals evolved twice within the tribe, in the subgenus *Lycoctonum* of *Aconitum* and in *Aconitella*. The fusion of the two most dorsal petals is uneven between species and between clades (which remains to be documented). However, it is whole in the 70 or so species of the *Consolida* group.

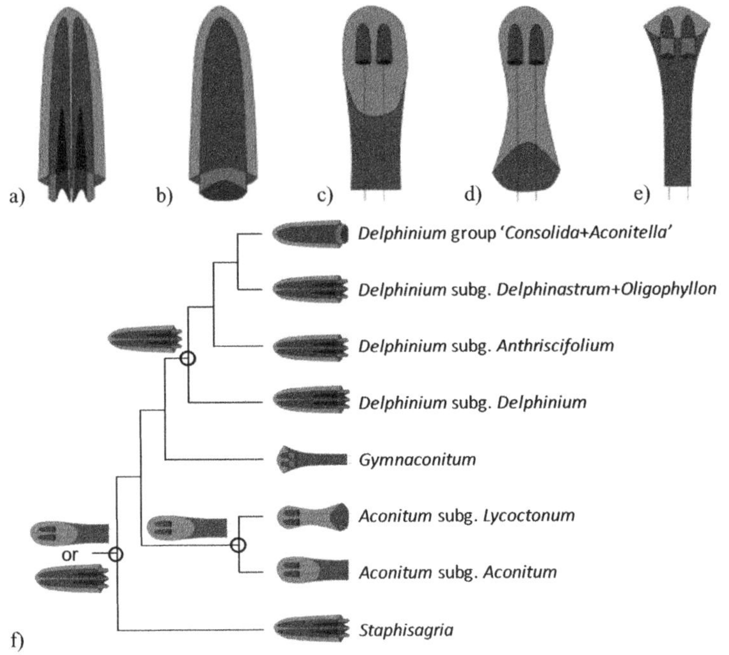

Figure 9.3. *Morphological diversity of the hyperorgan of the Delphinieae and distribution of the different morphological types on a phylogeny of the tribe. For a color version of this figure, see www.iste.co.uk/grandcolas/systematics.zip*

COMMENT ON FIGURE 9.3. – *a) Hyperorgan with two spurs included in a third. b) Hyperorgan with one spur resulting from the postgenital fusion of the primordia of the dorsal petals, included in a second. c–e) Hyperorgan with a pair of short spurs carried on a long pedicel, covered with a helmet (c), a cap (d) or a portion of the protruding dorsal sepal (e). f) Phylogeny of Delphinieae (Jabbour and Renner 2011, 2012b; Wang et al. 2013; Xiang et al. 2017). The morphological type of the hyperorgan is represented for three hypothetical ancestors (circled nodes).*

9.4. The study of synorganization in evolutionary biology

9.4.1. *Lessons learned from the synorganization study*

Understanding the nature of a hyperorgan, its ontogeny and function is an essential step in understanding the appearance of morphological innovation in the course of evolution, an innovation that confers a selective advantage to the organisms that are endowed with it. How do organs arrange themselves together, as well as sometimes merging, to form a structure with new emerging properties? What selective pressures are exerted on the synorganized structures over generations? How are architectural constraints circumvented in the physical association between structures with separate evolutionary histories, into a supra-structure characterized by its own evolutionary history? These questions are all avenues of interdisciplinary research in evolutionary botany.

9.4.2. *Scientometrics to measure the impact of the concept of synorganization in evolutionary biology*

Since the first use of the term "synorganization" in 1952, 12 research articles with this term in the title have been published (Google Scholar search with both "synorganisation" and "synorganization" spellings). The oldest (1956) is a zoology article, and the following 11 (period from 1980 to 2018) deal with botany. This number may seem low, but it should be kept in mind that this phenomenon is rarely documented, and that the concept can be, by its imprecise definition, difficult to manipulate. However, two articles have particularly marked the scientific community: Rudall and Bateman (2002), cited 192 times, and Endress (2015), cited 42 times.

9.4.3. *Synchronization, integration, co-adaptation, redundant concepts?*

Organisms are systems composed of more or less dependent or integrated elements. Modules are connected units (sometimes redundant), more strongly linked (or integrated) with each other than with other parts of the organic system (Olson and Miller 1958; Pigliucci 2003; Diggle 2014; Klingenberg 2014). Integration and modularity necessarily coexist: integration of the parts ensures the overall cohesion of the system and its resilience in the face of disruptions, while modularity allows functional specialization and the ability to adapt to internal or external changes of the organism (Armbruster *et al.* 2014).

Year	Author(s)	Title	Journal	Citations
1980	Froebe	Randlusterbildung und Synorganisation bei strahlenden Apiaceendolden	*Plant Syst. Evol.*	12
1981	Kunze	Morphogenese und Synorganisation des Bestaubungsapparates einiger Asclepiadaceen	*Beit. Biol. Pfl.*	36
1997	Douglas	The developmental basis of morphological diversification and synorganization in flowers of Conospermeae (*Stirlingia* and Conosperminae: Proteaceae)	*Int. J. Plant Sci.*	22
2002	Rudall and Bateman	Roles of synorganisation, zygomorphy and heterotopy in floral evolution: the gynostemium and labellum of orchids and other lilioid monocots	*Biol. Rev.*	192
2002	McMahon and Hufford	Developmental morphology and structural homology of corolla-androecium synorganization in the tribe Amorpheae (Fabaceae: Papilionoideae)	*Am. J. Bot.*	21
2003	Rudall *et al.*	Floral anatomy in *Dypsis* (Arecaceae–Areceae): a case of complex synorganization and stamen reduction	*Bot. J. Linn. Soc.*	28
2008	Claßen-Bockhoff and Heller	Floral synorganization and secondary pollen presentation in four Marantaceae from Costa Rica	*Int. J. Plant Sci.*	22
2010	Endress	Synorganisation without organ fusion in the flowers of *Geranium robertianum* (Geraniaceae) and its not so trivial obdiplostemony	*Ann. Bot.*	29
2012	Ley and Claßen-Bockhoff	Floral synorganization and its influence on mechanical isolation and autogamy in Marantaceae	*Bot. J. Linn. Soc.*	8
2015	Endress	Development and evolution of extreme synorganization in angiosperm flowers and diversity: a comparison of Apocynaceae and Orchidaceae	*Ann. Bot.*	42
2018	Nezha and Rezanezhad	The survey of synorganization of floral parts and pollinium and pollen germination and tube growth in *Calotropis procera* (Asclepiadoideae)	*J. Dev. Biol.*	0

Table 9.1. *Botanical research articles, whose title includes the term synorganisation or synorganization (Google Scholar search done on September 13, 2019)*

Is a synorganized structure necessarily integrated? Are the parts of a hyperorgan always comparable to modules? We do not think so. Even if modules could be defined in some synorganized structures (without each module necessarily corresponding to each part of the hyperorgan), the concepts of integration and synorganization are not interchangeable. While modularity is a specific feature of a system, synorganization can only be studied in an evolutionary framework and with reference to a sister group, where the homologous parts are not synorganized.

The term *synorganization* has been synonymized with the term *coaptation* by Vogel (1969) and Endress (1973). While the latter term first appeared in the early 18th century in a medical treatise (Stahl 1707), in zoology it is defined as the result of a perfect fit between two structures (Mayr and Provine 1998), or as the state of a system of parts that work together (Cuénot 1919). The masticatory apparatus of sea urchins, known as "Aristotle's lantern", is an example of coaptation (Savriama and Gerber 2018). Further research on the origin and different uses of the term *coaptation* is necessary to identify the singularities of this concept in relation to that of synorganization. Exchanges on this subject between zoologists and botanists will, in our opinion, provide the key to understanding how these concepts differ, and why their use remains very discipline-specific.

9.5. Conclusion

We continue our research on floral synorganization within the Delphinieae tribe, with a double objective: that of characterizing the morphological diversity of the hyperorgan on a large taxonomic scale (horizontal exploration), and of understanding the establishment of the hyperorgan during development (vertical exploration). In parallel, the morphogenesis of (non-synorganized) perianth elements in the genus *Nigella* (belonging to the sister group of Delphinieae, Zhai *et al.* (2019)) is analyzed by an evo-devo approach.

9.6. Acknowledgments

We thank the ISYEB (Institute of Systematics, Evolution and Biodiversity) and the MNHN (National Museum of Natural History, Paris) for their financial support, as well as the MNHN's electron microscopy technical platform and the team at the Paris Herbarium for their technical assistance. Florian Jabbour thanks the National Geographic Society (grant #9989-16) for funding a field mission to Malawi in 2019, and Catherine Damerval for discussions about synorganization.

9.7. References

APG (Angiosperm Phylogeny Group) (2016). An update of the angiosperm phylogeny group classification for the orders and families of flowering plants: APG IV. *Bot. J. Linn. Soc.*, 181(1), 1–20.

Armbruster, W.S., Pélabon, C., Bolstad, G.H., and Hansen, T.F. (2014). Integrated phenotypes: Understanding trait covariation in plants and animals. *Phil. Trans. R. Soc. B: Biol. Sci.*, 369(1649), 20130245.

Cuénot, L. (1919). La coaptation des fémurs antérieurs et de la tête chez les phasmes. *C.R. Acad. Sci.*, 169, 835–838.

Diggle, P.K. (2014). Modularity and intra-floral integration in metameric organisms: Plants are more than the sum of their parts. *Phil. Trans. R. Soc. B: Biol. Sci.*, 369(1649), 20130253.

Endress, P.K. (1973). Arils and aril-like structures in woody Ranales. *New Phytol.*, 72(5), 1159–1171.

Endress, P.K. (1990). Patterns of floral construction in ontogeny and phylogeny. *Biol. J. Linn. Soc.*, 39(2), 153–175.

Endress, P.K. (2010). Synorganisation without organ fusion in the flowers of *Geranium robertianum* (Geraniaceae) and its not so trivial obdiplostemony. *Ann. Bot.*, 106(5), 687–695.

Endress, P.K. (2016). Development and evolution of extreme synorganization in angiosperm flowers and diversity: A comparison of Apocynaceae and Orchidaceae. *Ann. Bot.*, 117(5), 749–767.

Jabbour, F. and Renner, S.S. (2011). *Consolida* and *Aconitella* are an annual clade of *Delphinium* (Ranunculaceae) that diversified in the Mediterranean basin and the Irano-Turanian region. *Taxon*, 60(4), 1029–1040.

Jabbour, F. and Renner, S.S. (2012a). Spurs in a spur: Perianth evolution in the Delphinieae (Ranunculaceae). *Int. J. Plant Sci.*, 173(9), 1036–1054.

Jabbour, F. and Renner, S.S. (2012b). A phylogeny of Delphinieae (Ranunculaceae) shows that *Aconitum* is nested within *Delphinium* and that Late Miocene transitions to long life cycles in the Himalayas and Southwest China coincide with bursts in diversification. *Mol. Phylogenet. Evol.*, 62(3), 928–942.

Klingenberg, C.P. (2014). Studying morphological integration and modularity at multiple levels: Concepts and analysis. *Phil. Trans. R. Soc. B: Biol. Sci.*, 369(1649), 20130249.

Ley, A.C. and Claßen-Bockhoff, R. (2012). Floral synorganization and its influence on mechanical isolation and autogamy in Marantaceae. *Bot. J. Linn. Soc.*, 168(3), 300–322.

Mayr, E. and Provine, W.B. (eds) (1998). *The Evolutionary Synthesis: Perspectives on the Unification of Biology*. Harvard University Press, Harvard.

Olson, E.C. and Miller, R.L. (1999). *Morphological Integration*. University of Chicago Press, Chicago.

Pigliucci, M. (2003). Phenotypic integration: Studying the ecology and evolution of complex phenotypes. *Ecol. Lett.*, 6(3), 265–272.

Remane, A. (1956). *Die Grundlagen des Natürlichen Systems, der Vergleichenden Anatomie und der Phylogenetik, Teil 1: Theoretische Morphologie und Systematik.* Akademische Verlagsgesellschaft Geest & Portig K.-G., Leipzig.

Ronse De Craene, L.P. (2010). *Floral Diagrams: An Aid to Understanding Flower Morphology and Evolution*. Cambridge University Press, Cambridge.

Salvini-Plawen, V.L.V. and Splechtna, H. (1979). Zur Homologie der Keimblätter. *J. Zool. Syst. Evol. Res.*, 17(1), 10–30.

Savriama, Y. and Gerber, S. (2018). Geometric morphometrics of nested symmetries unravels hierarchical inter- and intra-individual variation in biological shapes. *Sci. Rep.*, 8(1), 18055.

Specht, C.D. and Bartlett, M.E. (2009). Flower evolution: The origin and subsequent diversification of the angiosperm flower. *Annu. Rev. Ecol. Evol. Syst.*, 40, 217–243.

Stahl, G.E. (1707). *De vera diversitate corporis mixti et vivi, et utriusque peculiarium conditionum atque proprietatum necessaria discretione, demonstratio. Sumptibus litterisque*. Orphanotrophe, Halle.

Vogel, S. (1959). Organographie der bluten kaplandischer Ophrydeen mit bemerkungen zum koaptationsproblem. Teil I disinae und satyriinae. *Ahb. Akad. Wiss. Lit., Abh. Math.-Naturwiss. Kl.*, 6, 270–401.

Vogel, S. (1969). Über synorganisierte blütensporne bei einigen Orchideen. *Plant Syst. Evol.*, 116(1), 244–262.

Wagner, G.P. (1989). The biological homology concept. *Annu. Rev. Ecol. Syst.*, 20(1), 51–69.

Wang, W., Liu, Y., Yu, S.X., Gao, T.G., and Chen, Z.D. (2013). *Gymnaconitum*, a new genus of Ranunculaceae endemic to the Qinghai-Tibetan Plateau. *Taxon*, 62(4), 713–722.

Xiang, K.L., Aytaç, Z., Liu, Y., Espinosa, F., Jabbour, F., Byng, J.W., Zhang, C.F., Erst, A.S., and Wang, W. (2017). Recircumscription of *Delphinium* subg. *Delphinium* (Ranunculaceae) and implications for its biogeography. *Taxon*, 66(3), 554–566.

Zachos, F.E. and Hoßfeld, U. (2006). Adolf Remane (1898–1976) and his views on systematics, homology and the modern synthesis. *Theor. Biosci.*, 124(3–4), 335–348.

Zhai, W., Duan, X., Zhang, R., Guo, C., Li, L., Xu, G., Shan, H., Kong, H., and Ren, Y. (2019). Chloroplast genomic data provide new and robust insights into the phylogeny and evolution of the Ranunculaceae. *Mol. Phylogenet. Evol.*, 135, 12–21.

10

Identification of Interspecific Chromosomal Homologies: Chromosomal Microdissection and Chromosomal Painting in Antarctic Teleosts Nototheniidae

10.1. Introduction

10.1.1. *Homologies, painting and chromosomal microdissection*

Interspecific chromosomal homologies (ICH) refer to the correspondence between chromosomes of different species, inherited from a common ancestor (an inferred ancestral karyotype) and shared by all descendant species (the current observed karyotypes). This can be the establishment of a relationship of chromosomes in their entirety or a segment of these chromosomes between different species of a given taxon (Guan *et al.* 1994; Ried *et al.* 1998; Pinton *et al.* 2003; Blennow 2004; Ventura *et al.* 2009). Correspondences between paralogue segments within the same genome or genomes of individuals belonging to the same species are called *intraspecific chromosome homologies*.

Research in chromosomal homologies (CH) has often been carried out in humans (*Homo sapiens*) and their close relatives (Scherthan *et al.* 1994; Müller *et al.* 1999; Dumas *et al.* 2005; Picone *et al.* 2011), model species

Chapter written by Juliette AUVINET, Agnès DETTAÏ, Olivier CORITON, Catherine OZOUF-COSTAZ and Dominique HIGUET.

(Romanenko *et al.* 2007; Ventura *et al.* 2009) or domestic species and species of agronomic interest (Chaves *et al.* 2004; Nanda *et al.* 2011; Mandáková *et al.* 2017). In less frequent cases, they have been applied to groups of species evolving in natural environments (Graphodatsky *et al.* 2000; Ráb *et al.* 2008). The identification of these CHs has various purposes, including research of chromosomal anomalies (Breneman *et al.* 1993; Weimer *et al.* 1999; Blennow 2004) for therapeutic purposes (Guan *et al.* 1994; Ried *et al.* 1998; Arens *et al.* 2004). In a more fundamental framework, HC research within the same species makes it possible to identify duplications of paralogous genes, the establishment of multigenic families, as well as other large-scale rearrangements such as duplications of whole genomes (Jaillon *et al.* 2004). From the perspective of comparative studies (interspecific), knowledge of ICHs presents many interests, such as the identification of the origin of heteromorphic sexual pairs (Henning *et al.* 2008; Machado *et al.* 2011; Hirai *et al.* 2012), of supernumerary B chromosomes (Trifonov *et al.* 2002; Teruel *et al.* 2009; Voltolin *et al.* 2010), or the reconstruction of family relationships (Mandáková *et al.* 2017). These ICHs are then identifiable in a significant number of species (Ried *et al.* 1998; Ventura *et al.* 2009). They are thus used to trace structural inter-chromosomal rearrangements (fission, fusion, translocation types) (Ried *et al.* 1998; Vicari *et al.* 2015) in a logic of reconstructing evolutionary scenarios of specific (Romanenko *et al.* 2007; Stanyon *et al.* 2008; Picone *et al.* 2011) and/or chromosomal diversification (Yang *et al.* 2004; Ráb *et al.* 2008).

Different types of methodologies have been used in recent years to identify ICHs. These include banded marking (*GIEMSA* banding: G/R/C/GTG and *DAPI* banding), which enables the identification of pairs or chromosomal regions of interest based on the contrast of the segments obtained. It is a practical and inexpensive tool, offering the possibility of quickly establishing karyotype and CH, making it possible to reconstitute certain intra- or interchromosomal intra- or interspecific rearrangements. This technique is mainly used in mammals (Wienberg *et al.* 1992; Graphodatsky *et al.* 2000; Yang *et al.* 2004; Ventura *et al.* 2009) by marking the chromatid bands for which it offers good resolution (Wienberg *et al.* 1992; Dumas *et al.* 2005). Massive fusions have thus been detected in mouse populations (Ried *et al.* 1998), as well as other structural rearrangements in other rodents, such as gerbils (Dobigny *et al.* 2003), shrews and hamsters (Romanenko *et al.* 2007). In the same way, we can also cite the evidence of chromosomal rearrangements between cercopithecia and humans, or in birds and amphibians (Ried *et al.* 1998). However, banding does not always work systematically and is not always informative, even in taxa with highly reordered karyotypes (Chowdhary and Raudsepp 2001) (the morphology and size of chromosomes are too similar to distinguish them clearly one by one). For refined resolution at the gene scale, genotyping and sequencing whole genomes results in the reconstruction of large-scale molecular syntenies (Jaillon *et al.* 2004). The chromosomal segments studied, that are potentially subject to rearrangement, are thus more finely limited, even if the data generally only contain the gene regions, to the detriment of highly and moderately repeated sequences, which are difficult to assemble and reposition within the contiguous regions. The development of "next-generation" sequencing (NGS) technologies makes it possible to increase access to data that is sufficiently well-assembled and annotated for reliable and numerous reconstructions (Green 2001). Nevertheless, the difficulties of reconstructing large contigs and scaffolds (Green 2001) complicate the assembly and annotation of large genomes (Alkan *et al.* 2011).

Chromosome painting, used in comparative cytogenomics, is the most suitable and fastest tool to be implemented on a large number of species (Ried *et al.* 1998; Ventura *et al.* 2009) in order to detect ICHs and reconstruct the inter-chromosomal – not intra-chromosomal – rearrangements that have occurred during diversification. Thus, a *chromosome-specific painting probe* is a set of genomic DNA sequences from a single chromosome or chromosome segment of a species of interest (Laudicina and Mühlmann 2015; Vicari *et al.* 2015). After sampling, amplification and labeling of the DNA obtained, this probe is hybridized by fluorescence *in situ* hybridization (FISH) and paints the homologous chromosomes of the different species studied (Zoo-FISH). This technique is now developed and used in a wide range of taxa: dogs (Graphodatsky *et al.* 2000), wheat (Šafář *et al.* 2004), horses (Yang *et al.* 2004), mammals (Stanyon *et al.* 2008) and rodents (Ventura *et al.* 2009). Chromosomes (small chromosomes less than 10 Mb) can be sorted and separated according to their size and nitrogen base constitution by pulsed field gel electrophoresis (Schwartz and Cantor 1984) or by flow sorting (Gray *et al.* 1975a, 1975b; Monard 1998). Methodologies have also been developed to optimize the chromosome-specific signal by enriching these isolated chromosome products into unique sequences or by using a mixture of BAC (bacterial artificial chromosomes) selected to cover the entire desired segment. The number of repeat sequences present in the probe is then limited, thus reducing the possibility of hybridization to other chromosomes, or chromosome segments, than those targeted (Laudicina and Mühlmann 2015). The chromosome paintings that are already prepared, commercialized[1] and directly ready to use for FISH are probes that are very accessible and easy to apply. In addition to this enrichment in unique sequences, pre-hybridization of the DNA contained in the painting probe, with an excess amount of unlabeled genomic DNA enriched in repeated sequences (type Cot-1 DNA, also directly prepared and commercialized or competitor DNA), allows double-stranded re-matching of the repeated sequences and thus preferential hybridization of the unique sequences on the targeted chromosomes (Ried *et al.* 1998; Vicari *et al.* 2011; Swennenhuis *et al.* 2012; Trifonov *et al.* 2017).

Due to the genetic proximity between some mammals, it is possible to hybridize probes from heterologous paintings on the chromosomes of related species (dog–arctic/red fox (Graphodatsky *et al.* 2000), human–antelope (Chaves *et al.* 2004), human–elephant (Stanyon *et al.* 2008)). On the other hand, in taxa with a low degree of kinship with humans or domesticated species for which commercial probes exist (chickens, pigs, cattle), the genetic distance is too great for heterologous painting probes to hybridize on the targets. The separation of chromosomes one by one can also be compromised for karyotypes presenting a phenotypic homogeneity both in size and in base composition for several chromosomal pairs, thus decreasing the possibilities of using flow sorting (Pinton *et al.* 2003) for the constitution of paintings.

Chromosome microdissection then appears to be one of the only possible alternatives to prepare a "home-made" painting probe targeting a chromosomal region of interest in any species (Guan *et al.* 1994; Laudicina and Mühlmann 2015; Vicari *et al.* 2015). This methodology has been developed on drosophila polytene chromosomes (Hill and Watt 1977; Scalenghe *et al.* 1981), then in humans (Bates *et al.* 1986) and mice (Rohme *et al.* 1984). The first use of chromosome painting probes obtained from microdissected elements appeared in the early 1990s (Deng 1992; Kao 1993). This technique is now commonly used for a large number of organisms: chickens (Griffin *et al.* 1999), turtles (Ulsh *et al.* 2000), insects (Teruel *et al.* 2009), fish (Henning *et al.* 2008; Vicari *et al.* 2011) and plants (Stein *et al.* 1998; Matsunaga *et al.* 1999; Zhou *et al.* 1999; Kerk 2003). It

1 See, for example: www.cytocell.com, "Painting Probes LPP xxR/G".

consists of manually removing one or more chromosomes, chromosome arms or chromatid bands of the same pair from a chromosome preparation taken from a given individual, using a laser or glass micro-needle (Weimer *et al.* 1999; Vicari *et al.* 2015). Once amplified and labeled, this microdissection product is hybridized by FISH on the chromosomes of the species of interest. In rare cases, a single chromosome is sufficient to elaborate the painting probe (Guan *et al.* 1993; Viersbach *et al.* 1994; Christian *et al.* 1999; Griffin *et al.* 1999; Trifonov *et al.* 2002). However, it is usually necessary to obtain batches of several chromosomes of the same pair, constituting a minimum amount of microdissected product sufficient for good amplification. The identification of ICHs by microdissected painting probes has three prerequisites. It is necessary:

1) to be able to reliably and repeatably recognize/identify the pair of chromosomes to be microdissected within the different metaphases of the observed sample, due to the need to harvest several copies of this chromosome to make a probe (Zhou and Hu 2007; Vicari *et al.* 2015);

2) to study a group of species whose genomes are relatively conserved, in order to guarantee a sufficient identity of probe/target sequences and thus ensure their hybridization;

3) to have available and sufficient genomic DNA from the species from which the probe is derived, or from a genetically related species, in order to be able to develop competitor DNA or Cot-1 DNA to be prehybridized with the painting probe prior to FISH (Vicari *et al.* 2011; Swennenhuis *et al.* 2012; Trifonov *et al.* 2017).

10.1.2. *ICH research in Nototheniidae*

Over the last 40 million years, the cooling of the Southern Ocean and the establishment of the circumpolar current have led to the thermal isolation of the Antarctic continental shelf. The series of glacial and warming periods was the origin of the dynamics of the ice sheet, which was sometimes in progression (significant fragmentation of habitats in the presence of microniche refuges), sometimes in regression (re-circulation with the possibility of recolonization of vacant niches) (Kennett *et al.* 1985;

Kennett and Barker 1990; Stott *et al.* 1990). In this context, populations from the Nototheniidae family have diversified on the plateau to where they belong (Eastman 2000; Near *et al.* 2012; Lecointre *et al.* 2013). With 77% of the observed species diversity and more than 90% of the total biomass of the southern ichthyofauna, the Nototheniidae now constitute the most represented group of Antarctic teleosts (Eastman 2000; Detrich and Amemiya 2010; Detrich *et al.* 2010). Specific diversification has been accompanied by spectacular and rare intra- and interspecies chromosomal diversification (Pisano and Ozouf-Costaz 2000) among teleosts (Molina and de Freitas Bacurau 2006). Species belonging to the genus *Trematomus* have the highest chromosomal variability observed among all Nototheniidae, with a diploid number of chromosomes ranging from 24 to 58 (Ozouf-Costaz *et al.* 1991; Morescalchi 1992; Ghigliotti 2015). This suggests a very strong genomic plasticity, in particular, with numerous chromosomal rearrangements (fusions, fissions, inversions) that accompanied this recent radiation (dated at –9.1 Ma by Near *et al.* (2012), and –4.3 Ma by Colombo *et al.* (2015)).

In order to reconstruct an evolutionary scenario from these diversifications, it is necessary to establish ICHs between the different species of *Trematomus*, or, at a larger taxonomic scale, by considering the Nototheniidae as a whole. Unlike mammals, banding in teleosts does not allow reliable identification of chromosomal pairs, such as the comparison of karyotypes (Medrano *et al.* 1988). Genome sequences of the Nototheniidae *Notothenia coriiceps* and *Paracheanichthyis charcoti* are published and available, but, to date, are only partially assembled (Shin *et al.* 2014). Most chromosome pairs of species in this family are too similar to be clearly identified on the basis of size, morphology or DAPI contrast alone. The approach, by hybridization, of painting probes on the chromosomes of various Nototheniidae was first proposed by Ozouf-Costaz *et al.* (1997). Although initiated by Cocca *et al.* (2015) for the Y-differentiated sexual pair of *C. hamatus*, this method had never been carried out before in this group as part of a comparative approach.

We carried out a painting probe, starting from a specific chromosome obtained from the microdissection of the largest chromosome pair of *Trematomus pennellii*, probably originating from several fusions. It is one of the few pairs that can be easily identified by its very large size and submetacentric morphology. The choice of species to hybridize was made on species with very different karyotypes, including two *Trematomus*: *T. pennellii* (2n = 32) and *T. hansoni* (2n = 45/46, 48). Other representatives of the Nototheniidae family were included for comparison: *Notothenia coriiceps* (2n = 22) and *Dissostichus mawsoni* (2n = 48).

10.2. Materials and methods

10.2.1. *Materials*

The chromosome preparations were collected during the Antarctic campaigns of the IPEV (French Polar Institute Paul-Émile Victor) and NSF (US National Science Foundation) programs (Table 10.1): REVOLTA[2], ICOTA[3] and ICEFISH2[4]. The identification and karyotyping of the fish were carried out by Catherine Ozouf-Costaz, except for DM1, which was carried out by Guillaume Lecointre. The collection number per specimen is specified when it exists. The same chromosomal suspension of *T. pennellii*, specimen CE4313, was used for the microdissection step as well as for the FISH painting. The nomenclature of the SCAR atlas (Duhamel *et al.* 2014) was used for Table 10.1.

Subfamily	*Genus. species* (nb. diploids of chromosomes)	Field reference	Collection reference
Trematominae	*T. pennellii* (2n = 32)	CE 4313 (REVOLTA)	MNHN 1996-0319
	T. hansoni (2n = 46)	TA 646TRHA2 (ICOTA)	MNHN 1996-0307
Nototheniinae	*N. coriiceps* (2n = 22)	TA 415[2] (ICOTA)	MNHN 2008-1884
Dissostichinae	*D. mawsoni* (2n = 48)	DM1 (ICEFISH2)	No specimen of reference

Table 10.1. *Taxonomic sampling for chromosome suspensions*

2 Ecological resources and valorization by a long-term observatory in Terre Adélie, 2010–2014, Terre Adélie.

3 Coastal icthyology in Terre Adélie, 1996–2008, Terre Adélie.

4 *International Collaborative Expedition to collect and study Fish Indigenous to Sub-Antarctic Habitats*, 2004, Southern Ocean and South Atlantic.

10.2.2. *Methods*

10.2.2.1. *Microdissection and elaboration of the painting probe*

10.2.2.1.1. Chromosome preparations

The mitotic chromosomes of the different species of Nototheniidae were obtained by the method of Rey *et al.* (2015), from cell cultures of fragments (or samples) of cephalic kidney and spleen, hematopoietic organs in which numerous stem cells give rise to various cell lines (lymphocytes, lymphoid precursors, etc.). Cepalic kidney and spleen cells were briefly sieved and cultured in a bicarbonate-free Leibovitz L-15 medium enriched with L-glutamine, fetal calf serum, lectins (the concanavalin A mitogens and pokeweed) and antibiotics (antimycotic solution, Sigma-Aldrich) for a period of 6.5 days at 0 to +2°C. Colchicine was added to the media 6 hours before hypotonic shock (KCL 0.075 M, 1 hour at + 2°C). The resulting cell suspensions were fixed three times in a 3:1 absolute ethanol/ice acetic acid mixture and frozen in the field.

For spreading, the cell suspensions were thawed and centrifuged for 10 minutes at +4°C/1,500 rpm. The supernatant was removed and the pellet was resuspended in fresh fixative. FISH (1) and microdissection (2) were performed using a Pasteur pipette, by deposition on previously degreased Superfrost slides (1) or 24 x 60 mm slides (2) under steam in a water bath at 37°C to optimize the dispersion of metaphasic chromosomes.

The quality checking of the spreads was carried out using a phase contrast microscope (Zeiss, axioplan) to assess the quantity of correctly spread metaphases. The selected slides were stained with 4% Giemsa for 12 min.

10.2.2.1.2. Constitution of batches of microdissected homologous chromosomes

Microdissection of the largest pair of *T. pennellii* was performed using an inverted microscope (Axiovert 25, Zeiss) equipped with a micromanipulator joystick (Narishige). Chromosome isolation was obtained using glass needles made with a stretcher (Narishige). The protocols of Sefeirtova *et al.* (2013), Laudicina and Mühlmann (2015) and Vicari *et al.* (2015) guided us in our approach. Through calculation, we determined a minimum number of 30 homologous chromosomes of this largest pair of *T. pennellii* chromosomes to be microdissected for the constitution of the batches, taking a genome size of 1.15 pg (C value) as reference.

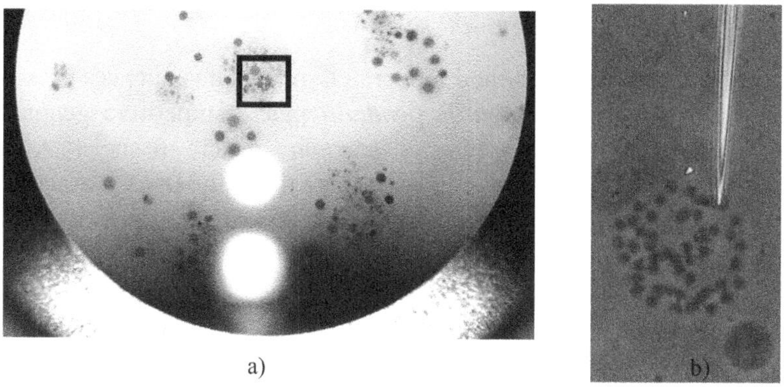

Figure 10.1. *Inverted microscope observation of two chromosome preparations for microdissection. For a color version of this figure, see www.iste.co.uk/grandcolas/systematics.zip*

COMMENT ON FIGURE 10.1.– a) Observation of the chromosomal spread of *T. pennellii* stained with GIEMSA at 400 magnification with a black framed metaphase. b) Observation of a metaphase of *O. niloticus* under an inverted microscope at 1,000 magnification. Glass needle placed under the objective near the largest pair of chromosomes.

10.2.2.1.3. Amplification, purification and quantification of products

Whole genome amplifications (WGA) of microdissected chromosome batches were performed using the protocols described by Gribble *et al.* (2004), Arneson *et al.* (2008) and Höckner *et al.* (2009). The GenomePlex kits of WGA4 and WGA3 (Sigma-Aldrich) were used following the supplier's protocol. The WGA4 kit was able to amplify an initial amount of DNA estimated between 4 and 6 pg, depending on the batches (Table 10.2, section 10.3.1). Amplification was verified by deposition on 1.5% agarose gel, and the amplification was purified by ethanol precipitation. Purified WGA4 product (10 ng) was reamplified with the WGA3 kit (minimum amount of material of 1 µg required for labeling).

At each stage of the development of the painting probe, samples and controls were accurately quantified using the Qubit 3.0 Fluorometer HS and BR (Invitrogen Corporation) kit.

10.2.2.2. *Development of competitor DNA*

In order to saturate the highly and moderately repeated sequences present in the painting probe of the large *T. pennellii* pair, competitive genomic DNA was prepared from *Trematomus hansoni* muscle tissue (TA646 TRHA2; see Table 10.1). The protocols used were those for DNA extraction: tissue lysis with cetyltrimethylammonium bromide (CTAB) and proteinase K, DNA recovery after treatment with isoamyl alcohol chloroform (IAC) and isopropanol pre-precipitation, rinsing with 70% ethanol. The DNA is fragmented by autoclaving and heat shock, then precipitated and resuspended in the hybridization buffer (50% formamide/2x SSC/10% dextran sulfate) at a final concentration of 8 µg/µl.

10.2.2.3. *Painting*

10.2.2.3.1. Probe labeling

The painting probes were indirectly labeled with biotin by random priming (Biotin-High Prime kit, Roche Diagnostics) according to the supplier's protocol, starting from an initial material quantity of 1 µg. The probes were precipitated and resuspended in a hybridization buffer (50% formamide/2x SSC/10% dextran sulfate) at a final concentration of 16 ng/µl.

10.2.2.3.2. FISH

To ensure the specificity of the painting signal, the FISHs were carried out under high stringency conditions. The biotinylated probes, denatured by heating at 85°C for 5 min in the presence of competitor DNA, were prehybridized for 2 hours at 37°C before being applied to the chromosome spreads, previously denatured at 72°C for a period of 10 seconds to 1 minute in a denaturation solution (70% formamide/2x SSC). The various washes and indirect detection using avidin-FITC (Fluorescein, Roche) were performed following the protocol of the "BAC probes" section of Bonillo *et al.* (2015).

10.2.2.3.3. Observation, capture and processing of images

The slides were analyzed under a microscope (Axioplan, Zeiss) equipped with a cooled B&W camera (Coolsnap Photometrics) and an epifluorescence lamp (XCite), coupled with karyotype/FISH-imaging software (CytoVision 3.93.2/Genus). A minimum of 30 observed metaphases per sample was required to ensure repeatability of results. The chromosomes of a metaphase were ordered according to the following criteria: decreasing size, morphological resemblance of homological pairs and DAPI contrast. Only after the chromosomes had been ordered were the painting signals superimposed on the computer-generated image.

10.3. Results

10.3.1. *Microdissection*

Two batches of 30 (no. 1 and no. 2) and one batch of 40 chromosomes (no. 3) were captured by chromosome microdissection (Table 10.2).

Batch no. 1 was directly amplified with the WGA4 kit, without using Proteinase K, in the first step of lysis and fragmentation of the genetic material. As it was not sufficiently amplified in WGA4, no amplification was detected after migration and detection on 1.5% of agarose gel. Quantification confirms a low concentration of batch no. 1 that is amplified in WGA4, less than 4 ng/µl, thus below the gel detection limit (Table 10.2).

Batches	Nb. chr.	Estimated initial quantity (pg)	Concentration after WGA4p (ng/µl)	Amplification rate after WGA4p	Concentration after WGA3p (ng/µl)	Purifying output (%)
No. 1	30	4.3	3.82	50,000	X	X
Positive control	X	10,000	38.6	270	X	X
No. 2	30	4.3	15.70	450,000	75.30	53.8
Positive control	X	10	21.50	370,000	151.00	73.7
No. 3	40	5.7	55	783,333	103.50	26.8
Positive control	X	10	153	540,000	170	58.7

Table 10.2. *Quantification of the different steps after amplifications and purifications to obtain the painting probe*

Batches no. 2 and 3 were amplified strictly following the protocol of the WGA4 kit supplier. The amplification was 9 and nearly 16 times greater than that of batch no. 1, respectively. These two batches were therefore re-amplified in WGA3. After various tests, we estimated that the amplification yields were significantly higher when the batches were systematically purified after amplification (removal of buffers, excess enzymes, adapters from WGA kits). Ethanol precipitation was chosen because, unlike other purification systems (columns), it makes it possible to choose the volume of resuspension of the material and thus, the final concentration before reamplification.

The positive control is a human DNA sample provided in the WGA4 (Sigma-Aldrich) kit.

Batch no. 3 gave the best result for the final quantity of material after WGA4, and then WGA3. In addition, no contamination was detected (see negative control, Figure 10.2). It was therefore retained for further processing. The genomic DNA fragments constituting the painting probe ranged from 100 bp to 1 kb, with a higher concentration of sequences between 200 and 400 bp (Figure 10.2), the optimal size of the probe fragments for access to their target during FISH (Vicari *et al.* 2015).

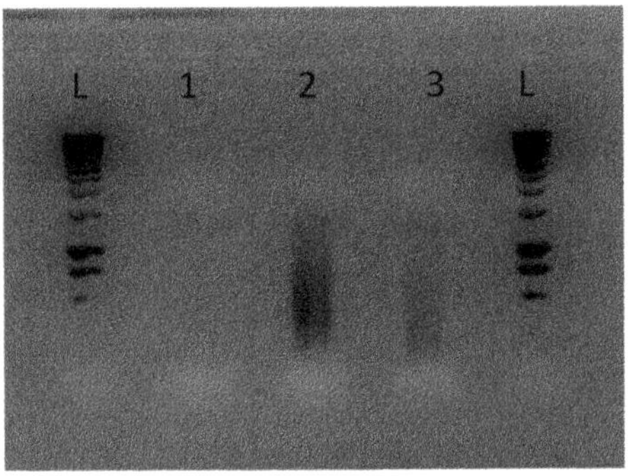

Figure 10.2. *Results of the amplification in WGA4 of batch no. 3 and controls by deposition on 1.5% of agarose gel*

COMMENT ON FIGURE 10.2.– Track 1: Negative control; Track 2: Positive control/human genomic DNA provided by the WGA4 (Sigma-Aldrich) kit (initial estimated amount = 10 pg/final dosed quantity = 5.4 g); Track 3: Microdissected and amplified product containing 40 chromosomes of the largest pair of T. pennellii (estimated initial amount = 8 pg/final dosed quantity = 4.7 g). L: BenchTop 100 bp molecular weight markers (Promega).

10.3.2. *Painting*

10.3.2.1. *Hybridizations of the chromosome-specific painting probe on chromosomes of the same specimen,* T. pennellii *(EC 4313) (positive control)*

The largest pair of chromosomes in *T. pennellii* is painted relatively homogeneously over its entire length (pair no. 1, Figure 10.3). This chromosome-specific signal is found in all observed metaphases. Other non-specific signals of the largest pair are also observed, more punctual and a minority compared to the other chromosome pairs. These ancillary signals may vary from one metaphase to another, but a very intense spot is systematically found in the pericentromeric region of the largest pair of acrocentric chromosomes of *T. pennellii* (pair no. 2, Figure 10.3).

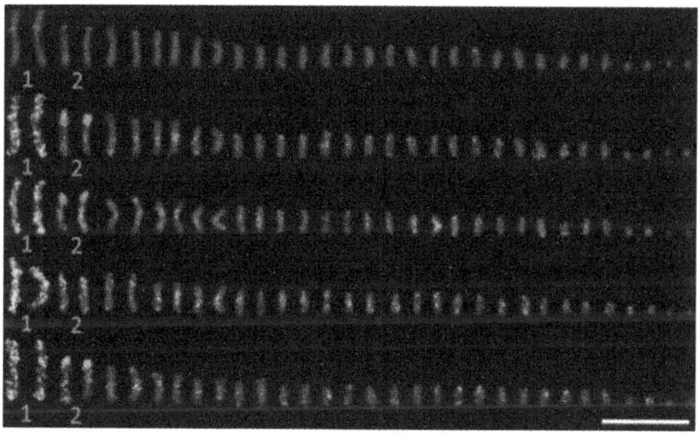

Figure 10.3. *Fluorescence* in situ *hybridization of the microdissected painting probe on* T. pennellii *chromosomes (adapted from Auvinet* et al. *(2020)). For a color version of this figure, see www.iste.co.uk/grandcolas/systematics.zip*

COMMENT ON FIGURE 10.3.– Presentation of a classification of the counterstained *T. pennellii* chromosomes on DAPI, followed by the four metaphases with ordered chromosomes, to which is added the painting signal. The painting probe is derived from the microdissection of the largest pair in size of *T. pennellii* (pair no. 1). It was amplified by the WGA method and indirectly labeled with biotin, then detected using avidin-FITC (fluorescein, green signal). Painted or easily identifiable pairs mentioned in the text are numbered (scale bar: 10 μm).

10.3.2.2. *Hybridations of the chromosome-specific painting probe on chromosomes of other species of Nototheniidae:* T. hansoni, N. coriiceps *and* D. mawsoni

When the painting probe is hybridized to the chromosomes of *T. hansoni* (2n = 46) and *D. mawsoni* (2n = 48), three pairs of acrocentric chromosomes with three chromosome arms are painted (numbered 3, 9 and 13, Figure 10.4a, and 9, 10 and 11, Figure 10.4b). This signal can be reproduced over all the pairs it covers. In a similar way to chromosome hybridization in *T. pennellii*, there are small and weak ancillary signals present in a dispersed manner on other pairs of chromosomes in both species.

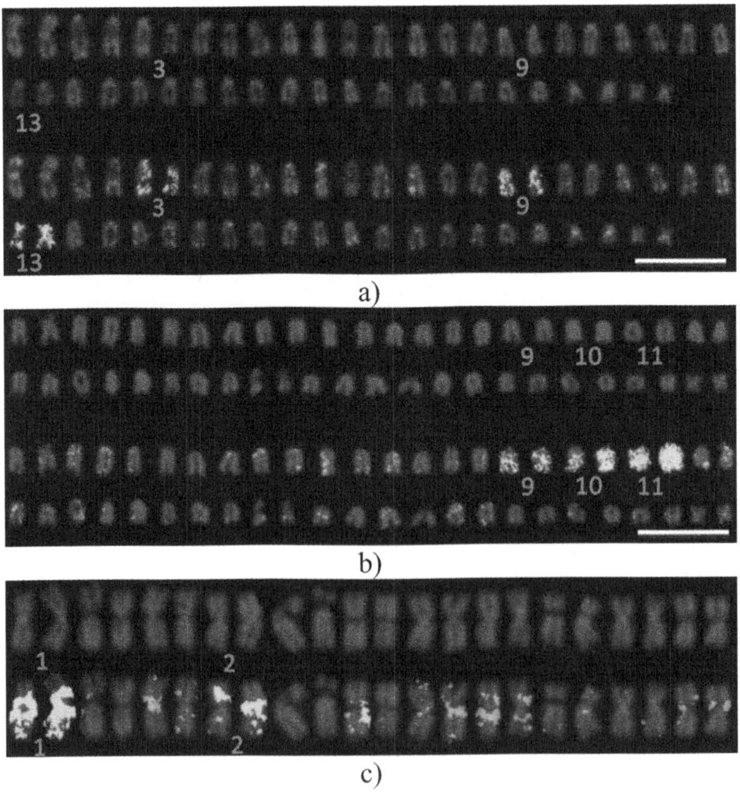

Figure 10.4. *Fluorescence* in situ *hybridization of the microdissected painting probe on the chromosomes of: a)* T. hansoni, *b)* D. mawsoni *and c)* N. coriiceps *((a) and (b) adapted from Auvinet* et al. *(2020)). For a color version of this figure, see www.iste.co.uk/grandcolas/systematics.zip*

COMMENT ON FIGURE 10.4.– For each of the figures, presentation of a classification of the counterstained chromosomes on DAPI, followed by the same image on which the painting signals appear. The painting probe is derived from the microdissection of chromosome pair no. 1 of *T. pennellii* (Figure 10.3). It was amplified using the WGA method and indirectly labeled in biotin, then detected using avidin-FITC (fluorescein, the green signal in the figure). The painted pairs mentioned in the text are numbered (scale bar: 10 μm).

The signals from the hybrid painting probe on the *N. coriiceps* chromosomes are more dispersed. Seven of the 11 pairs of chromosomes show an intense signal on at least one segment. Nevertheless, this signal covers a large part of the largest pair that can be easily identified by DAPI counterstaining (pair no. 1, Figure 10.3), as well as another smaller pair (pair no. 2, Figure 10.3). However, the signal does not cover the pairs in their entirety, as telomeric and subtelomeric regions are not marked. Fairly strong ancillary signals also cover several other pairs. They are concentrated in heterochromatic pericentromic regions (and, to a lesser extent, in telomeric regions), which are rich in satellite DNA.

10.4. Discussion

10.4.1. *Technical aspects developed and prospects for improvement of the painting signal*

The painting prepared from the microdissection of the largest pair of *T. pennellii* did produce signals specific to this largest pair, as well as ancillary signals detected on the other chromosome pairs. Even if there were sufficient metaphases on the chromosome preparations and the chromosomes were relatively well-spread within a metaphase to allow their sampling, there is always a risk of intrinsic contamination of batches through unwanted fragments of other chromosomes that have been picked up by the needle (Laudicina and Mühlmann 2015; Vicari *et al.* 2015). This risk is accentuated by the fact that multiple samples of the target chromosome are taken from different metaphases in order to form the batches (Weimer *et al.* 1999). In addition, the cytoplasmic residues are present, and act as a matrix binding the chromosomes together, which can lead to the synchronous detachment of several different chromosomes (Laudicina and Mühlmann 2015; Vicari *et al.* 2015).

One solution to minimize this risk and to maximize the number of potential detectable ICHs is to prepare painting probes from the microdissection of a small number of chromosomes (less than 5), or just one (Guan *et al.* 1993; Viersbach *et al.* 1994; Christian *et al.* 1999; Griffin *et al.* 1999; Trifonov *et al.* 2002). We have successfully performed a probe amplification, prepared from 15 chromosomes of the largest pair of *T. pennellii* or *T. hansoni*. This result shows that it is *a priori* possible to reduce the number of chromosomes for the constitution of paintings, and thus potentially eliminate part of the intrinsic contamination. In all these cases, including that of a painting preparation based on a single chromosome, it is, however, necessary to be able to clearly distinguish this chromosome for the sake of repeatability of the experiment. Sufficient size of the chromosomes, and, especially, the reliable identification of these pairs are thus strictly required. However, while some pairs for *T. pennellii* and *N. coriiceps* are recognizable on the basis of size, morphology and/or DAPI labeling (Figures 10.3 and 10.4c), it is not possible to recognize most pairs unequivocally (especially the small acrocentric pairs present in *T. pennellii*, *T. hansoni* and *D. mawsoni* (Figure 10.4a and b)). Therefore, this limits the number of possible candidate pairs for chromosomal microdissection, and consequently for the detection of ICHs by painting in this family.

The single-cell-type WGA4 techniques employed are based on the non-specific amplification of the collected product, using degenerated hexamers that act as primers (Vicari *et al.* 2015). They are optimized for equivocal amplification of the DNA content of a single human cell (7 pg) (Arneson *et al.* 2008; Hockner *et al.* 2009). It is therefore a method particularly suited to the amplification of microdissected chromosome batches (estimated at 4 pg for batches no. 1 and no. 2 and 6 pg for batch no. 3). However, there are high risks of contamination for microdissected samples by exogenous DNA because the WGA amplification step is sensitive (Laudicina and Mühlmann 2015). Contaminations due to the use of the WGA from batches no. 1 and no. 2 were detected during the development of the painting probe: negative controls consisting of sterile water, amplified with the WGA4 kit, produce low amplification during agarose gel deposition and a residual amount of contaminating DNA is detected during quantification. After having taken the necessary precautions for batch no. 3 (sterilization of the tools by UV exposure, no confinement in a fume hood when incorporating the products, opening of the tubes outside the handling area, etc.), we no longer detected any contamination. However, the sequencing of some of these products revealed a high proportion of contaminating sequences, mainly of human origin. Given the genetic distance between humans and Antarctic teleosts, these sequences could not prevent or bias the hybridization of the painting prepared on Nototheniidae chromosomes. On the other hand, the actual amount of nototheniid-specific probe was largely overestimated when dosing amplification products.

Despite the presence of competitor DNA during hybridization, the signal produced by the FISH of the painting was not just chromosome-specific. Scattered signals are indeed present on almost all other chromosomal pairs. While this "noise" signal is rather reduced in *T. hansoni* and *D. mawsoni*, it is more distinct in our positive control *T. pennellii* and in *N. coriiceps*. It is therefore difficult to clearly establish ICHs, because the chromosome-specific signal (corresponding to a majority of the hybridization of the unique sequences present in the probe) and the related repeated signals are unclear. The latter correspond to the highly and moderately repeated sequences present in greater or lesser quantities in the probe, and in all chromosomes: centromeric satellites, telomeric sequences and certain families of transposable elements.

Despite these signals being present on the other chromosome pairs, there is homogeneous coverage along the entire length of the largest pair of *T. pennellii*. While the distribution of the ancillary signals is rather fluctuant, the painting signal can be reproduced (observed on more than 70 different metaphases). These imperfections in the painting signal may be attributable to the method of chromosome fixation during their preparation (probable partial depurination of the DNA of the microdissected chromosomes by the acetic acid contained in the fixative, and thus degraded and less specific probe hybridization (Vicari *et al.* 2015)). These non-specific signals may also be due to amplification bias for repeated sequences, which are present in higher copy numbers compared to single sequences. This microdissected product is indeed a very large pair of chromosomes rich in heterochromatin, particularly at two constrictions (one being centromeric and one present in the median position in the large arm), and therefore more prone to contain accumulations of repeated sequences (Laudicina and Mühlmann 2015).

The painting approach of microdissected probes is interesting to identify these interspecific chromosomal homologies within the Nototheniidae, especially for some easily identifiable pairs. However, the use of these approaches would require technical improvements to make the signal easier to interpret. Chromosome preparation with the last binding in pure ethanol (without acetic acid) can be proposed to avoid depurination, and thus degradation of future probes. The development of Cot-1 DNA (rich in repeated sequences) could play the role of a "home made" competitor (Vicari *et al.* 2011; Swennenhuis *et al.* 2012; Trifonov *et al.* 2017). Improvement and optimization of the amplification and reamplification steps should be considered, in particular by reducing the number of chromosomes in each batch as much as possible and by taking precautions for human contaminations. Finally, the use of direct tagging of painting probes could significantly increase the resolution and power of the signal. In addition, the use of several different ("multicolor") paintings (Ried *et al.* 1998; Weimer *et al.* 2001) and their co-hybridization would allow a clearer distinction between the chromosome-specific signal (single fluorescence) and the repeated signal (superimposed fluorescence).

10.4.2. *The largest pair of chromosomes of* T. pennellii, *the product of two chromosomal fusions (roberstonian and tandem)*

In Nototheniidae, the use of painting probes is intended to enable the search for interspecific chromosomal homologies, so as to reconstruct the numerous rearrangements that occurred during their diversification (Pisano and Ozouf-Costaz 2000). With hybridization signals abundantly detected on the chromosomes of *T. hansoni*, *D. mawsoni* and *N. coriiceps*, the painting results obtained in the different species of this family suggest a good conservation of the sequences present in the largest pair of *T. pennellii* between the different species examined.

The ancestral karyotype of the Nototheniidae, like that of many teleosts, has been inferred at 2n = 48 acrocentric chromosomes (Jaillon *et al.* 2004; Kohn *et al.* 2006). This is also the case for karyotypes of the first diverging nototheniid lineages (Mazzei *et al.* 2006). The Nototheniidae present three types of chromosomal formulae to be linked with the degree of rearrangement of their karyotype:

1) 2n > 48 acrocentric chromosomes with probable fissions (*Trematomus nicolaï* 2n = 58);

2) 2n = 48 chromosomes that can be rearranged (by pericentric inversions) or not (acrocentric) (*D. mawsoni*);

3) 2n < 48 acrocentric, metacentric and submetacentric chromosomes, highly rearranged with pericentric fusions and/or inversions in greater or lesser proportions (*N. coriiceps* 2n = 24).

The last category concerns many species, notably *T. pennellii* (2n = 32), *T. hansoni* (2n = 46 here), as well as all species of the genus *Notothenia*, including *N. coriiceps* (2n = 22). This tendency to reduce the total number of chromosomes suggests that the rearrangements fixed during speciation would mostly be chromosomal fusions (Ozouf-Costaz 1997; Pisano and Ozouf-Costaz 2000). These can be of different types: centromeric/centromeric fusions also called *centric* or *robertsonian fusions*, or centromeric/telomeric fusions called *tandem fusions*.

The largest pair of chromosomes in *T. pennellii* appears to be homologous to three painted pairs of acrocentric chromosomes present in *T. hansoni* and *D. mawsoni*. The painted results thus suggest that this large pair of chromosomes of *T. pennellii* would result from a double fusion from three chromosomes inherited from the last common ancestor of the Nototheniidae, and shared by the progeny with a tandem fusion and a centric fusion (Figure 10.5). This result is consistent with the karyotype data of *T. pennellii* mentioned above (2n < 48).

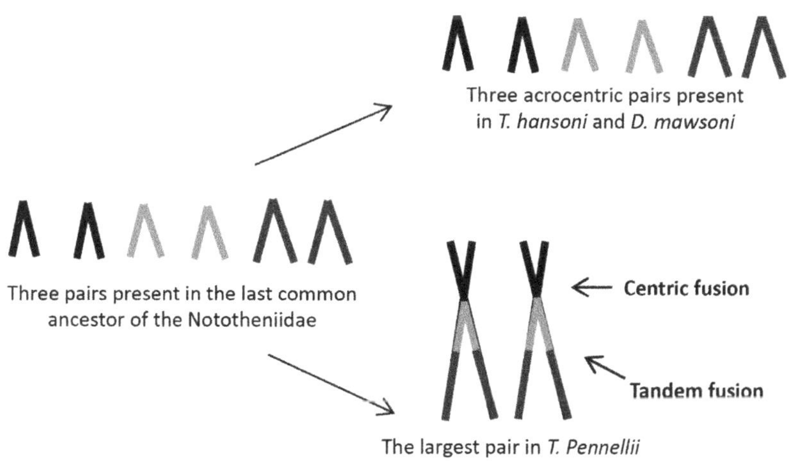

Figure 10.5. *Evolutionary scenario of chromosomal rearrangements leading to the largest pair of chromosomes in* T. pennellii. *For a color version of this figure, see www.iste.co.uk/grandcolas/systematics.zip*

COMMENT ON FIGURE 10.5.– The inferred plesiomorphic state is 2n = 48 acrocentric chromosomes present in the common ancestor of the Nototheniidae. Three pairs (painted by the painting probe) were preserved in the case of *T. hansoni* and *D. mawsoni*, and were dual-fused (centric and tandem fusions) in another case to give the largest pair of *T. pennellii* (painting probe).

Similarly, the results seem to indicate at least a partial homology of the largest chromosome pair of *T. pennellii* with that of *N. coriiceps*. This raises many questions concerning, for example, the homology of the chromosome fusions found in *N. coriiceps* (Nototheniinae) and *T. pennellii* (Trematominae). Are the same chromosome arms present in the common ancestor of Nototheniidae involved in the same fusions as the ones proposed within the Nototheniinae subfamily (one-to-one massive fusions (Amores *et al.* 2017))? On the other hand, this result raises the issue of the number of rearrangements that have occurred over time since the diversification of these species. The painting does not allow us to know if there has been one or more additional rearrangements after the pairs were fused. Other types of rearrangements could be detected using this tool, such as fission or translocations and inversions. One way to refine these painting results and increase the number of ICHs identified within the *Trematomus* is to no longer look at the scale of the chromosome as a whole, but as a finer unit defined by the chromosome arm.

10.5. Conclusion

Thanks to new NGS sequencing techniques that can generate a large amount of sequence data in a very short period of time, comparative genomics is booming. While whole genome sequencing is commonly performed for small (mostly unicellular) prokaryotic or eukaryotic genomes, as well as in model species, it remains a much more *ad hoc* process that is difficult to implement on whole lineages living in the wild (115 species are currently recorded in the Nototheniidae (Duhamel *et al.* 2014; Ghigliotti 2015) and only two species' whole genome has been sequenced, that of *N. coriiceps* and *P. charcoti*). On the other hand, the correct assembly and annotation of sequences to accurately establish ICHs requires time and precise expertise in this field.

The FISH approach appears to be an effective alternative to address these issues. By allowing the synthesis of probes by specific amplification of microdissected DNA, chromosomal painting makes it possible to overcome the lack of genomic data and commercialized probes in non-model species (Laudicina and Mühlmann 2015). The use of chromosomal painting thus opens up broad perspectives in the exploration of these genomes, in both fundamental and applied settings. Despite certain technical constraints inherent in the development of probes, it appears to be a powerful tool in molecular cytogenetics, being relatively easy and quick to implement (Ried *et al.* 1998; Vicari *et al.* 2015), and, above all, offering the possibility of examining a large number of species in the perspective of comparative studies.

10.6. References

Alkan, C., Sajjadian, S., and Eichler, E. (2011). Limitations of next-generation genome sequence assembly. *Nature Methods*, 8(1), 61–65.

Amores, A., Wilson, C.A., Allard, C.A.H., Detrich, H.W., and Postlethwait, J.H. (2017). Cold fusion: Massive karyotype evolution in the Antarctic Bullhead Notothen *Notothenia coriiceps*. *G: Genes|Genomes|Genetics*, 7(7), 2195–2207.

Arens, Y.H.J.M., Engelen, J.J.M., Govaerts, L.C.P., Van Ravenswaay, C.M., Loneus, W.H., Van Lent-Albrechts, J.C.M., Van der Blij-Philipsen, M., Hamers, A.J.H., and Schrander-Stumpel, C.T.R.M. (2004). Familial insertion (3;5)(q25.3;q22.1q31.3) with deletion or duplication of chromosome region 5q22.1-5q31.3 in ten unbalanced carriers. *American Journal of Medical Genetics*, 130A(2), 128–133.

Arneson, N., Hughes, S., Houlston, R., and Done, S. (2008). GenomePlex whole-genome amplification. *Cold Spring Harbor Protocols*, 2008(1).

Auvinet, J., Graça, P., Dettai, A., Amores, A., Postlethwait, J.H., Detrich, H.W. 3rd, Ozouf-Costaz, C., Coriton, O., and Higuet, D. (2020). Multiple independent chromosomal fusions accompanied the radiation of the Antarctic teleost genus *Trematomus* (Notothenioidei:Nototheniidae). *BMC Evolutionary Biology*, 20(1), 39.

Bates, G.P., Wainwright, B.J., Williamson, R., and Brown, S.D. (1986). Microdissection of and microcloning from the short arm of human chromosome 2. *Molecular and Cellular Biology*, 6(11), 3826–3830.

Blennow, E. (2004). Reverse painting highlights the origin of chromosome aberrations. *Chromosome Research*, 12(1), 25–33.

Bonillo, C., Coutanceau, J.P., D'Cotta, H., Ghigliotti, L., Ozouf-Costaz, C., and Pisano, E. (2015). Standard fluorescence *in situ* hybridization procedures. In *Fish Cytogenetic Techniques: Ray-Fin Fishes and Chondrichthyans*, Ozouf-Costaz, C., Pisano, E., Foresti, F., and Foresti de Almeida Toledo, L. (eds). CRC Press, Boca Raton.

Breneman, J.W., Ramsey, M.J., Lee, D.A., Eveleth, G.G., Minkler, J.L., and Tucker, J.D. (1993). The development of chromosome-specific composite DNA probes for the mouse and their application to chromosome painting. *Chromoma*, 102, 591–598.

Caron, F. (1988). Electrophorèse de l'ADN en champ pulsé. *International Journal of Medical Sciences*, 4(1), 46–49.

Chaves, R., Frönicke, L., Guedes-Pinto, H., and Wienberg, J. (2004). Multidirectional chromosome painting between the Hirola antelope (*Damaliscus hunteri*, Alcelaphini, Bovidae), sheep and human. *Chromosome Research*, 12(5), 495–503.

Chowdhary, B.P. and Raudsepp, T. (2001). Chromosome painting in farm, pet and wild animal species. *Methods in Cell Science*, 23, 37–55.

Christian, A.T., Garcia, H.E., and Tucker, J.D. (1999). PCR *in situ* followed by microdissection allows whole chromosome painting probes to be made from single microdissected chromosomes. *Mammalian Genome*, 10(6), 628–631.

Cocca, E., Petraccioli, A., Morescalchi, M.A., Odierna, G., and Capriglione, T. (2015). Laser microdissection-based analysis of the Y sex chromosome of the Antarctic fish *Chionodraco hamatus* (Notothenioidei, Channichthyidae). *Comparative Cytogenetics*, 9(1), 1–15.

Colombo, M., Damerau, M., Hanel, R., Salzburger, W., and Matschiner, M. (2015). Diversity and disparity through time in the adaptive radiation of Antarctic notothenioid fishes. *Journal of Evolutionary Biology*, 28(2), 376–394.

Deng, H.-X., Yoshiura, K., Dirks, R.W., Harada, N., Hirota, T., Tsukamoto, K., Jinno, Y., and Niikawa, N. (1992). Chromosome-band-specific painting: Chromosome *in situ* suppression hybridization using PCR products from a microdissected chromosome band as a probe pool. *Human Genetics*, 89(1), 13–17.

Detrich, H.W. and Amemiya, C.T. (2010). Antarctic Notothenioid fishes: Genomic resources and strategies for analyzing an adaptive radiation. *Integrative and Comparative Biology*, 50(6), 1009–1017.

Detrich, H.W., Stuart, A., Schoenborn, M., Parker, S.K., Methé, B.A., and Amemiya, C.T. (2010). Genome enablement of the notothenioidei: Genome size estimates from 11 species and BAC libraries from 2 representative taxa. *Journal of Experimental Zoology Part B: Molecular and Developmental Evolution*, 314B(5), 369–381.

Dobigny, G., Ozouf-Costaz, C., Bonillo, C., and Volobouev, V. (2003). Evolution of rRNA gene clusters and telomeric repeats during explosive genome repatterning in *Taterillus* X (Rodentia, Gerbillinae). *Cytogenetic and Genome Research*, 103(1–2), 94–103.

Dobigny, G., Aniskin, V., Granjon, L., Cornette, R., and Volobouev, V. (2005). Recent radiation in West African *Taterillus* (Rodentia, Gerbillinae): The concerted role of chromosome and climatic changes. *Heredity*, 95(5), 358–368.

Duhamel, G., Hulley, P.A., Causse, R., Koubbi, P., Vacchi, M., Pruvost, P., Vigetta, S., and Irisson, J.O. (2014). Part 7: Biogegographic patterns of fish. In *Biogeographic Atlas of the Southern Ocean*, De Broyer, C., Koubbi, P., Griffiths, H.J., Raymond, B., Udekem d'Acoz, C., Van de Putte, A.P., Danis, B., David, B., Grant, S., Gutt, J., Held, C., Hosie, G., Huettmann, F., Post, A., and Ropert-Coudert, Y. (eds). SCAR, Cambridge, 328–352.

Dumas, F., Bigoni, F., Stone, G., Sineo, L., and Stanyon, R. (2005). Mapping genomic rearrangements in titi monkeys by chromosome flow sorting and multidirectional *in-situ* hybridization. *Chromosome Research*, 13(1), 85–96.

Eastman, J. (2000). Fishes on the Antarctic continental shelf: Evolution of a marine species flock? *Journal of Fish Biology*, 57, 84–102.

Farce, M.-H. and Perez, J.-M. (2000). *Génétique moléculaire : principes et application aux populations animales*. Inra, Paris.

Ghigliotti, L., Cheng, C.C.H., Ozouf-Costaz, C., Vacchi, M., and Pisano, E. (2015). Cytogenetic diversity of notothenioid fish from the Ross sea: Historical overview and updates. *Hydrobiologia*, 761(1), 373–396.

Graphodatsky, A.S., Yang, F., O'Brien, P.C.M., Serdukova, N., Milne, B.S., Trifonov, V., and Ferguson-Smith, M.A. (2000). A comparative chromosome map of the Arctic fox, red fox and dog defined by chromosome painting and high resolution G-banding. *Chromosome Research*, 8(3), 253–263.

Gray, J.W., Carrano, A.V., Moore, D.H., Steinmetz, L.L., Minkler, J., Mayall, B.H., Mendelsohn, M.L., and Van Dilla, M.A. (1975a). High-speed quantitative karyotyping by flow microfluorometry. *Clinical Chemistry*, 21(9), 1258–1262.

Gray, J.W., Carrano, A.V., Steinmetz, L.L., Van Dilla, M.A., Moore, D.H., Mayall, B.H., and Mendelsohn, M.L. (1975b). Chromosome measurement and sorting by flow systems. *Proceedings of the National Academy of Sciences*, 72(4), 1231–1234.

Green, E.D. (2001). Strategies for the systematic sequencing of complex genomes. *Nature Reviews Genetics*, 2(8), 573–583.

Gribble, S., Ng, B.L., Prigmore, E., Burford, D.C., and Carter, N.P. (2004). Chromosome paints from single copies of chromosomes. *Chromosome Research*, 12, 143–151.

Griffin, D.K., Haberman, F., Masabanda, J., O'Brien, P., Bagga, M., Sazanov, A., Smith, J., Burt, D.W., Ferguson-Smith, M., and Wienberg, J. (1999). Micro- and macrochromosome paints generated by flow cytometry and microdissection: Tools for mapping the chicken genome. *Cytogenetic and Genome Research*, 87(3–4), 278–281.

Guan, X.-Y., Trent, J.M., and Meltzer, P.S. (1993). Generation of band-specific painting probes from a single microdissected chromosome. *Human Molecular Genetics*, 2(8), 1117–1121.

Guan, X.Y., Meltzer, P.S., and Trent, J.M. (1994). Rapid generation of whole chromosome painting probes (WCPs) by chromosome microdissection. *Genomics*, 22, 101–107.

Henning, F., Trifonov, V., and Foresti de Almeida-Toledo, L. (2008). Use of chromosome microdissection in fish molecular cytogenetics. *Genetics and Molecular Biology*, 31(1), 279–283.

Hill, R.J. and Watt, F. (1977). Chromosomal proteins of *Drosophila melanogaster* and an approach for their localisation on polytene chromosomes. *Chromosoma*, 63(1), 57–78.

Hirai, H., Hirai, Y., and LoVerde, P.T. (2012). Evolution of sex chromosomes ZW of Schistosoma mansoni inferred from chromosome paint and BAC mapping analyses. *Parasitology International*, 61(4), 684–689.

Hockner, M., Erdel, M., Spreiz, A., Utermann, G., and Kotzot, D. (2009). Whole genome amplification from mircodissected chromosomes. *Cytogenetic and Genome Research*, 125, 98–102.

Jaillon, O., Aury, J.-M., Brunet, F. *et al.* (2004). Genome duplication in the teleost fish *Tetraodon nigroviridis* reveals the early vertebrate proto-karyotype. *Nature*, 431(7011), 946–957.

Kao, F.-T. (1993). Genes and genomes: Microdissection and microcloning of human chromosome regions in genome and genetic disease analysis. *BioEssays*, 15(2), 141–146.

Kennett, J.P. and Barker, P.F. (1990). Latest Cretaceous to Cenozoic climate and oceanographic development in the Weddell Sea: An ocean drilling perspective. *Proceedings of the Ocean Drilling Project Science Results*, 113, 937–960.

Kennett, J.P., Keller, G., and Srinavasan, M.S. (1985). Miocene planktonic foraminiferai biogeography and paieoceauographic development of the indo-Pacific region. *Geol. Soc. Am. Mem*, 163, 197–236.

Kerk, N.M. (2003). Laser capture microdissection of cells from plant tissues. *Plant Physiology*, 132(1), 27–35.

Kohn, M., Hogel, J., Vogel, W., Minich, P., Kehrersawatzki, H., Graves, J., and Hameister, H. (2006). Reconstruction of a 450-My-old ancestral vertebrate protokaryotype. *Trends in Genetics*, 22(4), 203–210.

Laudicina, A. and Mühlmann, M.C. (2015). Fish chromosome microdissection and its utilization in the synthesis of fluorescence *in situ* hybridization probes. In *Fish Cytogenetic Techniques: Ray-Fin Fishes and Chondrichthyans*, Ozouf-Costaz, C., Pisano, E., Foresti, F., and Foresti de Almeida Toledo, L. (eds). CRC Press, Boca Raton.

Lecointre, G., Améziane, N., Boisselier, M.-C., Bonillo, C., Busson, F., Causse, R., Chenuil, A., Couloux, A., Coutanceau, J.-P., Cruaud, C., d'Udekem d'Acoz, C., De Ridder, C., Denys, G., Dettaï, A., Duhamel, G., Eléaume, M., Féral, J.-P., Gallut, C., Havermans, C., Held, C., Hemery, L., Lautrédou, A.-C., Martin, P., Ozouf-Costaz, C., Pierrat, B., Pruvost, P., Puillandre, N., Samadi, S., Saucède, T., Schubart, C., and David, B. (2013). Is the species flock concept operational? The Antarctic shelf case. *PLOS One*, 8(8), e68787.

Machado, T.C., Pansonato-Alves, J.C., Pucci, M.B., Nogaroto, V., Almeida, M.C.. Oliveira, C., Foresti, F., Bertollo, L.A.C., Moreira-Filho, O., Artoni, R.F., and Vicari, M.R. (2011). Chromosomal painting and ZW sex chromosomes differentiation in *Characidium* (Characiformes, Crenuchidae). *BMC genetics*, 12(1), 65.

Mandáková, T., Pouch, M., Harmanová, K., Zhan, S.H., Mayrose, I., and Lysak, M.A. (2017). Multispeed genome diploidization and diversification after an ancient allopolyploidization. *Molecular Ecology*, 26(22), 6445–6462.

Matsunaga, S., Kawano, S., Michimoto, T., Higashiyama, T., Nakao, S., Sakai, A., and Kuroiwa, T. (1999). Semi-automatic laser beam microdissection of the Y chromosome and analysis of Y chromosome DNA in a dioecious plant, *Silene latifolia*. *Plant and Cell Physiology*, 40(1), 60–68.

Mazzei, F., Ghigliotti, L., Lecointre, G., Ozouf-Costaz, C., Coutanceau, J.-P., Detrich, W., and Pisano, E. (2006). Karyotypes of basal lineages in notothenioid fishes: The genus *Bovichtus*. *Polar Biology*, 29(12), 1071–1076.

Medrano, L., Bernardi, G., Couturier, J., Dutrillaux, B., and Bernardi, G. (1988). Chromosome banding and genome compartimentalization in fishes. *Chromosoma*, 96, 178–183.

Molina, W.F. and de Freitas Bacurau, T.O. (2006). Structural and numerical chromosome diversification in marine Perciformes (Priacanthidae and Gerreidae). *Cytologia*, 71(3), 237–242.

Monard, S.P. (1998). Chromosome sorting and analysis by FACS. In *Flow Cytometry Protocols*, Jaroszeski, M.J. and Heller, R. (eds). Humana Press, Totowa.

Morescalchi, A., Pisano, E., Stanyon, R., and Morescalchi, M.A. (1992). Cytotaxonomy of antarctic teleosts of the *Pagothenia/Trematomus* complex (Nototheniidae, Perciformes). *Polar Biology*, 12(6–7), 553–558.

Müller, S., Stanyon, R., O'brien, P.C.M., Ferguson-Smith, M.A., Plesker, R., and Wienberg, J. (1999). Defining the ancestral karyotype of all primates by multidirectional chromosome painting between tree shrews, lemurs and humans. *Chromosoma*, 108(6), 393–400.

Nanda, I., Benisch, P., Fetting, D., Haaf, T., and Schmid, M. (2011). Synteny conservation of chicken macrochromosomes 1–10 in different avian lineages revealed by cross-species chromosome painting. *Cytogenetic and Genome Research*, 132(3), 165–181.

Near, T.J., Dornburg, A., Kuhn, K.L., Eastman, J.T., Pennington, J.N., Patarnello, T., Zane, L., Fernández, D.A., and Jones, C.D. (2012). Ancient climate change, antifreeze, and the evolutionary diversification of Antarctic fishes. *Proceedings of the National Academy of Sciences*, 109(9), 3434–3439.

Ozouf-Costaz, C., Hureau, J.-C., and Beaunier, M. (1991). Chromosome studies on fish of the suborder Notothenioidei collected in the Weddell sea during EPOS 3 cruise. *Cybium*, 15(4), 271–289.

Ozouf-Costaz, C., Pisano, E., Thaeron, C., and Hureau, J.-C. (1997). Antarctic fish chromosome banding: Significance for evolutionary studies. *Cybium*, 21(4), 399–409.

Pansonato-Alves, J.C., Vicari, M.R., Oliveira, C., and Foresti, F. (2011). Chromosomal diversification in populations of *Characidium* cf. *gomesi* (Teleostei, Crenuchidae). *Journal of Fish Biology*, 78(1), 183–194.

Picone, B., Masters, J., Silvestro, D., Sineo, L., and DelPero, M. (2011). A phylogenetic analysis of human syntenies revealed by chromosome painting in Euarchontoglires orders. *Journal of Mammalian Evolution*, 18(2), 131–146.

Pinton, A., Ducos, A., and Yerle, M. (2003). Chromosomal rearrangements in cattle and pigs revealed by chromosome microdissection and chromosome painting. *Genetics Selection Evolution*, 35(6), 685–696.

Pisano, E. and Ozouf-Costaz, C. (2000). Chromosome change and the evolution in the Antarctic fish suborder Notothenioidei. *Antarctic Science*, 12(3).

Ráb, P., Rábová, M., Pereira, C.S., Collares-Pereira, M.J., and Pelikánová, Š. (2008). Chromosome studies of European cyprinid fishes: Interspecific homology of leuciscine cytotaxonomic marker – the largest subtelocentric chromosome pair as revealed by cross-species painting. *Chromosome Research*, 16(6), 863–873.

Rey, O., d'Hont, A., Coutanceau, J.P., Pisano, E., Chilmonczyk, S., and Ozouf-Costaz, C. (2015). Cephalic kidney and spleen cell culture in Antarctic teleosts. In *Fish Cytogenetic Techniques: Ray-Fin Fishes and Chondrichthyans*, Ozouf-Costaz, C., Pisano, E., Foresti, F., and Foresti de Almeida Toledo, L. (eds). CRC Press, Boca Raton.

Ried, T., Schröck, E., Ning, Y., and Wienberg, J. (1998). Chromosome painting: A useful art. *Human Molecular Genetics*, 7(10), 1619–1626.

Rohme, D., Fox, H., Herrmann, B., Frischauf, A.-M., Edstrom, J.-E., Mains, P., Silver, L.M., and Lehrach, H. (1984). Molecular clones of the mouse t complex derived from microdissected metaphase chromosomes. *Cell*, 36, 783–788.

Romanenko, S.A., Volobouev, V.T., Perelman, P.L., Lebedev, V.S., Serdukova, N.A., Trifonov, V.A., Biltueva, L.S., Nie, W., O'Brien, P.C.M., Bulatova, N.S., Ferguson-Smith, M.A., Yang F., and Graphodatsky, A.S. (2007). Karyotype evolution and phylogenetic relationships of hamsters (Cricetidae, Muroidea, Rodentia) inferred from chromosomal painting and banding comparison. *Chromosome Research*, 15(3), 283–297.

Šafář, J., Bartoš, J., Janda, J., Bellec, A., Kubaláková, M., Valárik, M., Pateyron, S., Weiserová, J., Tušková, R., Číhalíkova, J., Vrána, J., Šimková, H., Faivre-Rampant, P., Sourdille, P., Caboche, M., Bernand, M., Doležel, J., and Chalhoub, B. (2004). Dissecting large and complex genomes: Flow sorting and BAC cloning of individual chromosomes from bread wheat. *The Plant Journal*, 39(6), 960–968.

Scalenghe, F., Turco, E., Edström, J.E., Pirrotta, V., and Melli, M. (1981). Microdissection and cloning of DNA from a specific region of *Drosophila melanogaster* polytene chromosomes. *Chromosoma*, 82(2), 205–216.

Scherthan, H., Cremer, T., Arnason, U., Weier, H.-U., Lima-de-Faria, A., and Fronicke, L. (1994). Comparative chromosome painting discloses homologuous segments in distantly related mammals. *Nature Genetics*, 6, 342–347.

Schwartz, D.C. and Cantor, C.R. (1984). Separation of yeast chromosome-sized DNAs by pulsed field gradient gel electrophoresis. *Cell*, 37, 67–75.

Seifertova, E., Zimmerman, L.B., Gilchrist, M.J., Macha, J., Kubickova, S., Cernohorska, H., Zarsky, V., Owens, N.D.L., Sesay, A.K., Tlapakova T., and Krylov, V. (2013). Efficient high-throughput sequencing of a laser microdissected chromosome arm. *BMC Genomics*, 14(1), 357.

Shin, S.C., Ahn, D.H., Kim, S.J., Pyo, C.W., Lee, H., Kim,M.-K., Lee, J., Lee, J.E., Detrich, H.W., Postlethwait, J.H., Edwards, D., Lee, S.G., Lee J.H., and Park, H. (2014). The genome sequence of the Antarctic bullhead notothen reveals evolutionary adaptations to a cold environment. *Genome Biology*, 15(9), 1.

Stanyon, R. and Stone, G. (2008). Phylogenomic analysis by chromosome sorting and painting. In *Phylogenomics*, Murphy, W.J. (ed.). Humana Press, Totowa, 13–29.

Stein, N., Ponelies, N., Musket, T., McMullen, M., and Weber, G. (1998). Chromosome micro-dissection and region-specific libraries from pachytene chromosomes of maize (*Zea mays* L.). *Plant Journal*, 13(2), 281.

Stott, L.D., Kennett, J.P., Shackleton, N.J., and Corfield, R.M. (1990). The evolution of antarctic surface waters during the paleogene: Inferences from the stable isotopic composition of planktonic foraminifers, ODP Leg 1131. *Proceedings of the Ocean Drilling Project Science Results*, 113, 849–863.

Swennenhuis, J.F., Foulk, B., Coumans, F.A.W., and Terstappen, L.W.M.M. (2012). Construction of repeat-free fluorescence *in situ* hybridization probes. *Nucleic Acids Research*, 40(3), e20.

Telenius, H., Ponder, B.A.J., Tunnacliffe, A., Carter, N.P., Behmel, A., Ferguson-Smith, M.A., Nordenskjöld, M., Pfragner, R. and Ponder, B.A. (1992). Cytogenetic analysis by chromosome painting using DOP-PCR amplified flow-sorted chromosomes. *Genes, Chromosomes and Cancer*, 4(3), 257–263.

Teruel, M., Cabrero, J., Montiel, EE., Acosta, M.J., Sánchez, A., and Camacho, J.P.M. (2009). Microdissection and chromosome painting of X and B chromosomes in *Locusta migratoria. Chromosome Research*, 17(1), 11–18.

Trifonov, V.A., Perelman, P.L., Kawada, S.-I., Iwasa, M.A., Oda, S.-I., and Graphodatsky, A.S. (2002). Complex structure of B-chromosomes in two mammalian species: *Apodemus peninsulae* (Rodentia) and *Nyctereutes procyonoides* (Carnivora). *Chromosome Research*, 10(2), 109–116.

Trifonov, V.A., Vorobieva, N.V., Serdyukova, N.A., and Rens, W. (2017). FISH with and without COT1 DNA. In *Fluorescence* In Situ *Hybridization (FISH)*, Liehr, T. (ed.). Springer, Berlin, Heidelberg, 123–133.

Ulsh, B.A., Mühlmann-Díaz, M.C., Whicker, F.W., Hinton, T.G., Congdon, J.D., and Bedford, J.S. (2000). Chromosome translocations in turtles: A biomarker in a sentinel animal for ecological dosimetry. *Radiation Research*, 153(6), 752–759.

Ventura, K., O'Brien, P.C.M., Yonenaga-Yassuda, Y., and Ferguson-Smith, M.A. (2009). Chromosome homologies of the highly rearranged karyotypes of four *Akodon* species (Rodentia, Cricetidae) resolved by reciprocal chromosome painting: The evolution of the lowest diploid number in rodents. *Chromosome Research*, 17(8), 1063–1078.

Vicari, M.R., de Mello Pistune, H.F., Castro, J.P., de Almeida, M.C., Bertollo, L.A.C., Moreira-Filho, O., Camacho, J.P.M., and Artoni, R.F. (2011). New insights on the origin of B chromosomes in *Astyanax scabripinnis* obtained by chromosome painting and FISH. *Genetica*, 139(8), 1073–1081.

Vicari, M.R., Pansonato-Alves, J.C., and Foresti, F. (2015). Microdissection and chromosome painting. In *Fish Cytogenetic Techniques: Ray-Fin Fishes and Chondrichthyans*, Ozouf-Costaz, C., Pisano, E., Foresti, F., and Foresti de Almeida Toledo, L. (eds). CRC Press, Boca Raton.

Viersbach, R., Schwanitz, G., and Nöthen, M.M. (1994). Delineation of marker chromosomes by reverse chromosome painting using only a small number of DOP-PCR amplified microdissected chromosomes. *Human Genetics*, 93(6), 663–667.

Voltolin, T.A., Laudicina, A., Senhorini, J.A., Bortolozzi, J., Oliveira, C., Foresti, F., and Porto-Foresti, F. (2010). Origin and molecular organization of supernumerary chromosomes of *Prochilodus lineatus* (Characiformes, Prochilodontidae) obtained by DNA probes. *Genetica*, 138(11–12), 1133–1139.

Weimer, J., Kiechle, M., Senger, G., Wiedemann, U., Ovens-Raeder, A., Schuierer, S., Kautza, M., Siebert, R. and Arnold, N. (1999). An easy and reliable procedure of microdissection technique for the analysis of chromosomal breakpoints and marker chromosomes. *Chromosome Research*, 7, 355–362.

Weimer, J., Koehler, M.R., Wiedemann, U., Attermeyer, P., Jacobsen, A., Karow, D., Kiechle, M., Jonat, W., and Arnold, N. (2001). Highly comprehensive karyotype analysis by a combination of spectral karyotyping (SKY), microdissection, and reverse painting (SKY-MD). *Chromosome Research*, 9, 395–402.

Wienberg, J., Stanyon, R., Jauch, A., and Cremer, T. (1992). Homologies in human and *Macaca fuscata* chromosomes revealed by *in situ* suppression hybridization with human chromosome specific DNA libraries. *Chromosoma*, 101, 265–270.

Yang, F., Fu, B., O'Brien, P.C.M., Nie, W., Ryder, O.A., and Ferguson-Smith, M.A. (2004). Refined genome-wide comparative map of the domestic horse, donkey and human based on cross-species chromosome painting: Insight into the occasional fertility of mules. *Chromosome Research*, 12, 65–76.

Zhou, R.-N. and Hu, Z.-M. (2007). The development of chromosome microdissection and microcloning technique and its applications in genomic research. *Current Genomics*, 8, 67–72.

Zhou, Y., Hu, Z., Dang, B., Wang, H., Deng, X., Wang, L., and Chen, Z. (1999). Microdissection and microcloning of rye (*Secale cereale* L.) chromosome 1R. *Chromosoma*, 108, 250–255.

List of Authors

Juliette AUVINET
Institut de Systématique,
Évolution, Biodiversité (ISYEB)
CNRS-MNHN
Sorbonne University
and
EPHE
and
University of the Antilles
Paris
France

Michel BAGUETTE
Institut de Systématique,
Évolution, Biodiversité (ISYEB)
CNRS-MNHN
Sorbonne University
and
EPHE
and
University of the Antilles
Paris
France

Philippe BASTIN
Institut Pasteur
INSERM
Paris
France

Joris BERTRAND
Laboratoire Génome et
Développement des Plantes
University of Perpignan
France

Isabelle BOUCHART-DUFAY
French National Museum of
Natural History
Paris
France

Estefanía CALVO ALVAREZ
Institut Pasteur
INSERM
Paris
France

Mathilde CARPENTIER
Institut de Systématique,
Évolution, Biodiversité (ISYEB)
CNRS-MNHN
Sorbonne University
and
EPHE
and
University of the Antilles
Paris
France

Jacques CHOMILIER
IMPMC
CNRS-MNHN
Sorbonne University
IRD
Paris
France

Olivier CORITON
INRAE
Rennes
FRANCE

Agnès DETTAÏ
Institut de Systématique,
Évolution, Biodiversité (ISYEB)
CNRS-MNHN
Sorbonne University
and
EPHE
and
University of the Antilles
Paris
France

Samuel FRACHON
Institut de Systématique,
Évolution, Biodiversité (ISYEB)
CNRS-MNHN
Sorbonne University
and
EPHE
and
University of the Antilles
Paris
France

Myriam GAUDEUL
Institut de Systématique,
Évolution, Biodiversité (ISYEB)
CNRS-MNHN
Sorbonne University
and
EPHE
and
University of the Antilles
Paris
France

Sylvain GERBER
Institut de Systématique,
Évolution, Biodiversité (ISYEB)
CNRS-MNHN
Sorbonne University
and
EPHE
and
University of the Antilles
Paris
France

Delphine GEY
2AD
French National Museum of Natural History
CNRS
Paris
France

Philippe GRANDCOLAS
Institut de Systématique, Évolution, Biodiversité (ISYEB)
CNRS-MNHN
Sorbonne University
and
EPHE
and
University of the Antilles
Paris
France

Dominique HIGUET
Institut de Systématique, Évolution, Biodiversité (ISYEB)
CNRS-MNHN
Sorbonne University
and
EPHE
and
University of the Antilles
Paris
France

Florian JABBOUR
Institut de Systématique, Évolution, Biodiversité (ISYEB)
CNRS-MNHN
Sorbonne University
and
EPHE
and
University of the Antilles
Paris
France

Nina JOFFARD
CEFE
CNRS
Montpellier
France

Line LE GALL
Institut de Systématique, Évolution, Biodiversité (ISYEB)
CNRS-MNHN
Sorbonne University
and
EPHE
and
University of the Antilles
Paris
France

Marie-Christine MAUREL
Institut de Systématique,
Évolution, Biodiversité (ISYEB)
CNRS-MNHN
Sorbonne University
and
EPHE
and
University of the Antilles
Paris
France

Antoine MOREL
Institut de Systématique,
Évolution, Biodiversité (ISYEB)
CNRS-MNHN
Sorbonne University
and
EPHE
and
University of the Antilles
Paris
France

Catherine OZOUF-COSTAZ
CNRS
Paris
France

Tony ROBILLARD
Institut de Systématique,
Évolution, Biodiversité (ISYEB)
CNRS-MNHN
Sorbonne University
and
EPHE
and
University of the Antilles
Paris
France

Germinal ROUHAN
Institut de Systématique,
Évolution, Biodiversité (ISYEB)
CNRS-MNHN
Sorbonne University
and
EPHE
and
University of the Antilles
Paris
France

Florence ROUSSEAU
Institut de Systématique,
Évolution, Biodiversité (ISYEB)
CNRS-MNHN
Sorbonne University
and
EPHE
and
University of the Antilles
Paris
France

Yoland SAVRIAMA
Max Planck Institute for Evolutionary Biology
Plön
Germany

Bertrand SCHATZ
CEFE
CNRS
Montpellier
France

Amir YASSIN
Institut de Systématique, Évolution, Biodiversité (ISYEB)
CNRS-MNHN
Sorbonne University
and
EPHE
and
University of the Antilles
Paris
France

Julie ZALKO
Institut de Systématique, Évolution, Biodiversité (ISYEB)
CNRS-MNHN
Sorbonne University
and
EPHE
and
University of the Antilles
Paris
France

Index

Other titles from

ISTE

in

Biology and Biomedical Engineering

2020

CAZEAU Cyrille
Foot Surgery Viewed Through the Prism of Comparative Anatomy: From Normal to Useful
(Comparative Anatomy and Posture of Animal and Human – Volume 1)

DUJON Bernard, PELLETIER Georges
Trajectories of Genetics

2019

BRAND Gérard
Discovering Odors

BUIS Roger
Biology and Mathematics: History and Challenges

2016

CLERC Maureen, BOUGRAIN Laurent, LOTTE Fabien
Brain-Computer Interfaces 1: Foundations and Methods
Brain-Computer Interfaces 2: Technology and Applications

FURGER Christophe
Live Cell Assays: From Research to Health and Regulatory Applications

2015

CLARYSSE Patrick, FRIBOULET Denis
Multi-modality Cardiac Imaging: Processing and Analysis

2014

CHEZE Laurence
Kinematic Analysis of Human Movement

CLARYSSE Patrick, FRIBOULET Denis
Multi-modality Cardiac Imaging: Processing and Analysis

DAO Tien Tuan, HO BA THO Marie-Christine
Biomechanics of the Musculoskeletal: Modeling of Data Uncertainty and Knowledge

FANET Hervé
Medical Imaging Based on Magnetic Fields and Ultrasounds

FARINAS DEL CERRO Luis, INOUE Katsumi
Logical Modeling of Biological Systems

MIGONNEY Véronique
Biomaterials

TEBBANI Sihem, LOPES Filipa, FILALI Rayen, DUMUR Didier, PAREAU Dominique
CO_2 Biofixation by Microalgae: Modeling, Estimation and Control

Printed and bound by CPI Group (UK) Ltd, Croydon, CR0 4YY

15/09/2026

14962345-0002